Wissenschaftsethik und Technikfolgenbeurteilung
Band 4

Schriftenreihe der Europäischen Akademie zur Erforschung
von Folgen wissenschaftlich-technischer Entwicklungen
Bad Neuenahr-Ahrweiler GmbH
herausgegeben von Carl Friedrich Gethmann

Springer

Berlin
Heidelberg
New York
Barcelona
Hong Kong
London
Milan
Paris
Singapore
Tokyo

John Grin · Armin Grunwald (Eds.)

Vision Assessment: Shaping Technology in 21st Century Society

Towards a Repertoire for Technology Assessment

 Springer

Reihenherausgeber

Professor Dr. Carl Friedrich Gethmann
Europäische Akademie zur Erforschung von Folgen
wissenschaftlich-technischer Entwicklungen
Bad Neuenahr-Ahrweiler GmbH
Postfach 1460, D-53459 Bad Neuenahr-Ahrweiler

Bandherausgeber

Dr. John Grin
Afdeling Politicologie
Universiteit van Amsterdam
OZ Achterburgwal 237, NL - 1012 DL Amsterdam

Priv.-Doz. Dr. Armin Grunwald
Institut für Technikfolgenabschätzung und Systemanalyse
Forschungszentrum Karlsruhe
Postfach 3640, D - 76021 Karlsruhe

Redaktion

Dagmar Uhl, M. A.
Europäische Akademie GmbH
Postfach 1460, D - 53459 Bad Neuenahr-Ahrweiler

Library of Congress Cataloging-in-Publication Data applied for
Die Deutsche Bibliothek - Cip-Einheitsaufnahme
Vision assessment: shaping technology in 21" century society / John Grin, Armin Grunwald (eds.).
Berlin; Heidelberg; New York; Barcelona; Hongkong; London; Milan; Paris; Singapore; Tokyo: Springer, 2000
(Wissenschaftsethik und Technikfolgenbeurteilung; Bd. 4)
ISBN-13: 978-3-642-64092-6

ISBN-13: 978-3-642-64092-6   e-ISBN-13: 978-3-642-59702-2
DOI: 10.1007/978-3-642-59702-2

© Springer-Verlag Berlin Heidelberg 2000
Softcover reprint of the hardcover 1st edition 2000

Coverdesign: de'blik, Berlin; Production: ProduServ GmbH Verlagsservice, Berlin
Typesetting: Camera-ready by editors
                    SPIN:10747743   62/3020SC - 5 4 3 2 1 0

European Academy
for the Study of Consequences
of Scientific and Technological Advance
Bad Neuenahr-Ahrweiler GmbH

The European Academy

The *European Academy Bad Neuenahr-Ahrweiler GmbH* is concerned with the scientific study of consequences of scientific and
technological advance for the individual and social life and for the
natural environment. The European Academy intends to contribute
to a rational way of society of dealing with the consequences of
scientific and technological developments. This aim is mainly realised in the development of recommendations for options to act,
from the point of view of long-term societal acceptance. The work
of the European Academy mostly takes place in temporary interdisciplinary project groups, whose members are recognized scientists from European universities. Overarching issues, e.g. from the
fields of Technology Assessment or Ethics of Science, are dealt with
by staff of the European Academy.

The Series

The Series „Ethics of Science and Technology Assessment" serves
to publish the results of the European Academy's work. It is published by the Academy's director. Besides the final results of the
project groups the series includes volumes on general questions of
ethics of science and technology assessment as well as other mono-
graphic studies.

Foreword

Life – in science just as in society in general – develops around visions, projects and unintended events that offer both challenges for existing co-operations and opportunities for new ones. The particular project from which this book results has its origins with two events. In October 1996, at the bi-annual Conference of the European Association for the Study of Science and Technology (EASST) in Biele-feld, John Grin bought two just appeared books by two German teams on the role of visions in socio-technical developments. One came from Meinolf Dierkes and his colleagues at the Science Centre Berlin, the other from Peter Mambrey and others at the GMD National Research Centre for Information Technology in Sankt Augustin. These studies touched upon one of Grin's key interests: how can we, in a reflexive and critical-reflective way, come to terms with the long term future? The concept of visions draws our attention to the fact that this is not just a naive question of how to sketch blueprints for the next fifty years, but a matter of criti-cally judging and adapting of the visions that simply *are* already guiding devel-opments in societal sectors into ones that more people like better.

The second event occurred a few months later, when the chief-of-operations of the Royal Dutch Air Force invited John Grin to give a 'provocative lecture on technological and conceptual planning of the airforces' at an International Collo-quium on *The Future of Airpower*, to be held late November 1997. The invitation of giving such a lecture in front of senior officers from all NATO countries was provocative enough to just accept it, but having been out of the field of military technology and security for some years this immediately created the necessity for an intensive literature survey. While preparing this lecture some of the problems that occurred reminded him of Mambrey *c.s.*'s work. Their 'discourse analysis' of the visions they found among their colleagues revealed a similar bias in the vi-sions that dominate 21st century airpower concepts.

That was the start of a session John Grin proposed for the 1998 EASST confer-ence in Lisbon, on 'Expectations, Technology, the 21st Century and Technology Assessment.' Rob Reuzel and Gert Jan van der Wilt, whose both critical and clini-cally oriented work on medical technology assessment had been a ground for frequent exchange since long, fortunately agreed to join the session. Their work, relying on sound medical knowledge, ethics and technology assessment method-ology, added the sharp view implied by any attempt to justice to both established socio-technological development and, sometimes fundamental, criticism of it.

Parallel, another chain of events had developed in Bad Neuenahr-Ahrweiler, Germany, where the European Academy for the Study of Consequences of Scien-tific and Technological Advance had settled itself. As a young, European-oriented institute it wishes to contribute to TA in Europe both through innovative work by its own staff and through exchange with colleagues from all over Europe. That made Armin Grunwald, vice-director of the European Academy and also inter-ested in long term planning and TA, to propose a paper to the forum of the 1998 EASST Conference. In spring 1998, the Academy invited John Grin to discuss with them the methodical guide on interactive TA he had recently co-authored. At that occasion, John Grin asked Armin Grunwald to contribute to the session at the EASST Conference, because his paper added to the others a both praxeological and fundamental reflection, betraying Grunwald's departure from planning theory, ethics and epistemology, on the tension between long-term planning and public legitimacy.

The project got real impetus in Lisbon. We experienced not only that we share an interest in research on long term orientations that is fundamental and critical as well as empirical and constructive. Even more stimulating, we discovered that a variety of themes were running through our respective papers, in ways that were both fascinating *and* too intricate to explore them, during that session, in any other way than very superficially. Our idea to publish a volume where we could explore these matters further turned into a project to be realised in the recently started book series of the European Academy.

It appeared that another Academy researcher, Michael Decker, could enrich the book with a chapter on robotics applications in health care, which is the only *ex ante* analysis in the book, a proposal on how to start a TA-project on a new technology, of which little understanding exists outside expert circles.

Intensive exchanges along all 5 × 4 conceivable axes did the rest to elaborate the commonality of our contributions, both by writing the first and final chapters and by adapting the original conference papers.

June, 1999   John Grin
Amsterdam/Bad Neuenahr   Armin Grunwald

Contents

IV   Conclusions

Outline

Michael Decker, John Grin, Armin Grunwald, Peter Mambrey, Rob Reuzel, August Tepper, Gert Jan van der Wilt

In this book, we take up the challenge of the role of technology in the first half of the 21st century, as we are supposedly entering the 'post-industrial', 'trans-national', 'post-modern' (and so on) society. We will assume that one way to shape socio-technological systems is through the visions that guide their development. The idea is *not* to create such visions bolt from the blue. Rather, the assumption is that visions exist already in most societal sectors, that these visions tend to reproduce the ways in which these sectors have developed hitherto, and that a critical discussion of these visions is a prerequisite for changing the course of development. We will ask how we can critically assess and construct visions on the ways in which technology and social problems are going to relate to each other so as to support critical discursive moments in existing recursive practices.

We thereby attempt to shed some light not only on existing visions, but also on thinking about guiding long-term societal development. Is it possible, indeed, to provide some orientation to long-term development in a way that *i)* contributes to meeting challenges like the need for sustainable development; while *ii)* escaping the risk of authoritarian blueprints and *iii)* ensures it public legitimacy.

Dealing with these matters, this volume draws upon insights from technology studies, policy studies, epistemology, sociology and ethics. It is to contribute to the recent stream of literature in technology studies on 'shaping technology', taking into account the 'co-evolution of technology and society'. It connects to that technology assessment literature that emphasises TA's pro-active role and its contribution to political judgement. It uses those insights from policy planning and epistemology that may help to reconcile long-term planning and public legitimacy. It considers socio-technological systems from a broader point of view, taking into account both macro-sociological and ethical analysis.

A book like this, with a variety of chapters that rather than discussing a series of research questions one by one, each in their own way touch upon a set of common themes, cannot be introduced in any linear way. John Grin introduces these themes, suggesting questions and exploring answers, and relates this to the chapters that follow. It is organised around the question how we may, somehow, shape the future, escaping the risk of falling into authoritarian utopias but simultaneously addressing some of the serious problems that the 20th century has brought us. It discusses the possibility that processes of political judgement, in which guiding visions are being discussed and transformed, may be at the core of the answer. But, as it is emphasised, pluriform judgement goes beyond the widespread habit of integrating into a particular, dominant vision those elements of alternative visions that may help to make that dominant vision more robust.

What we are rather looking for are ways to reach what Hans Georg Gadamer has designated a 'fusion of horizons.'

The chapter thus sets the stage for the remainder of this volume, in which we explore these issues further around the central question to what extent and in what ways TA can contribute to shaping the future through critical and constructive assessment of existing visions and the assumptions underlying them. However, it does not give a systematic overview of the chapters that follow. This we do below, not only to provide the reader with some additional guidance, but also because this roots these chapters into the various research programmes from which they originate, and which have been indicated above.

Peter Mambrey and August Tepper report on findings how metaphors and guiding visions were used and could be used to assist the design of Information and Communications Technologies (ICT). The work is based on research done in the project "Metaphors in Computer Science. Potentials and Risks". The research took place at GMD – German National Research Center for Information Technology, where a large-scale research program on assisting computers was begun ten years ago. The goals of the program were to study and develop the principles and methods for assisting systems, and to demonstrate assistance properties and capabilities in various prototype systems.

This specific project's aim was to investigate the methodological use of guiding visions and metaphors for technical design and for preventively oriented technology assessment and to suggest adequate tools. The project combined different research arenas which were previously unconnected. It combined technology assessment (future assessment), genesis of technology (role of *Leitbilder* in social systems), and linguistics (construction and analysis of metaphors). Two case studies about the role of *Leitbilder* and metaphors in the development of the typewriter and the personal computer demonstrated the scope of the approach as well as the predictive qualities. A survey on how members of a research and development organisation for applied information systems used and constructed metaphors was done.

Additionally to the language oriented assessment approach, the discourses of the prototypes of the assistance computer within the organisation had been analysed. Only by analysing and by contrasting the selected metaphors used by others one can identify important things included or excluded from the design process. The interpretation and contrastation of the written materials of the discourses between the scientists gave us insight into the concepts and ideas which were behind the metaphors. The *Leitbild*-centered analysis and the discourse-centred analysis both cannot predict or discover utopia but they are useful means and instruments to open perspectives on certain aspects, highlighting them, and obscure others, downplaying them. They are means for communication and enable discourses on technologies which are under construction or even not yet developed.

Rob Reuzel and Gert Jan van der Wilt present a case in the area of health technology assessment (HTA). This is a tool for judging the value or merit of medical interventions in view of subsequent decision-making. Criteria of merit and standards of performance derive from visions, expectations, or perspectives with respect to a technology. If one particular vision, expectation, or perspective dominates technology assessment, and is not shared by all persons involved, this assessment suffers from normative bias. Such was the case with cochlear implantation in prelingually deaf children, which is elaborated on in the chapter "Technology assessment in the health care area". With respect to this example, the authors answer three questions: (a) did HTA help to uncover underlying visions, expectations, and perspectives, (b) should it, and (c) how can it?

The answer to the first question reads: no, HTA did not uncover underlying visions, expectations, and perspectives on cochlear implantation; on the contrary, it reinforced the medical-technical perspective on deafness as a handicap to be eradicated, which contradicts the deaf perspective on deafness as a cultural feature of a linguistic minority. The perspective on deafness as a handicap perfectly corresponds to the professional vision on health as something to be normalised by experts.

The answer to the second question reads: yes, HTA should uncover underlying visions, expectations, and perspectives, for if it aims at valuing technology, it should be able to account for the validity of value standards used, that is, explain for whom this technology has value, how much, and why.

Finally, with respect to the third question, Reuzel and van der Wilt argue that interactive technology assessment procedures are promising. In order to provide an answer to the problems met in current mainstream HTA, such a procedure should feature: eliciting multiple perspectives and value frameworks, meeting various information needs, democratic consensus building, and shared decision-making.

John Grin starts his article on TA's role behind new airpower concepts with a suspicion concerning the so-called Revolution in Political and Military Affairs that permeates much of current writing on defence postures and military strategy. Part of the rationale is, as one document puts it, that the American public "[expects] quick victory and abhors unnecessary casualties" and "reserves the right to reconsider their support should any of these conditions not be met". To understand their argument, it is furthermore necessary to point out that they foresee, in the 21st century, so-called "collisions of war forms". While the United States and other western countries will increasingly fight according to third wave concepts, their adversaries in many cases will be following second wave or even first wave prescriptions.

This revolution is claimed to be a very profound one. Yet, Grin's claim is, it is not revolutionary at all in the sense that it reflects basic assumptions that are characteristic for High Modernity. The chapter investigates this more closely for technological and conceptual development in airpower. It starts with outlining the emergence and substance of the dominant vision guiding airpower planning from 1903 through its canonisation in two recent, influential airpower publications. The basic assumptions that return in subsequent visions are that airpower's basic attributes (speed, acceleration, range, carrying power) must be maximally exploited; that it is thus, ultimately, possible to overcome operational limitations; and that in this way there is a direct and substantial contribution of technology to strategic objectives.

Subsequently, it is argued that the 'revolutionary' concepts now being proposed for the 21st century, basically reflect the same horizon. This raises the suspicion that they may suffer from similar shortcomings as those brought forward between 1975 and 1989 against then existing airpower technology and strategic and operational concepts. The technology approaches used in technology assessments investigating such criticism are then used to assess concepts for 21st century airpower. The chapter ends with a discussion of how and where alternative concepts may be identified and elaborated, and how TA may contribute to that exercise.

While the previous chapters deal with identifying, assessing and reconstructing visions in particular sectors, the next part of the book turns to a different perspective. It attempts to understand the problems and opportunities for technology assessment to contribute to shaping the future in a way that does not run into the traps of a variety of well-known dilemmas. Armin Grunwald's article deals with the problem of maintaining long-term agendas in technology and environmental policy. Modern societies are, on the one hand, confronted with strong requirements for reliable *long-term orientation* of technology policy, environmental policy and science policy. On the other hand, these policies must obviously be based on some kind of acceptance. Acceptance, however, is depending on contingent contextual conditions and may vary rather rapidly dependent on risk and benefit perception and chance events. Technology and environmental policy, therefore, are seemingly running into a dilemma: without acceptance at all they must fail because forms of resistance will raise, but to base them mainly or completely on acceptance cannot safeguard the long-term orientations required.

The approach presented to deal with this dilemma starts from the observation that if it is acknowledged that *without* respect to acceptance reliable long-term planning cannot be reached in a democratic and pluralistic society, and that relying *only* on acceptance does not allow to follow long-term plans. The only way remaining is to *shift the level of acceptance* required. The question to be answered is *what elements* of technology policy must really be based on acceptance. The method used to attain an answer is reconstructing the underlying rationality standards of society as basis for deriving criteria for the *acceptability* of technology. The level of acceptance required is, thus, shifted from the acceptance of the factual technology to the acceptance of the rationally justified criteria and procedures – from substantial acceptance to a *procedural one*. This shift allows to combine the acceptance-orientation of a pluralistic democracy and the long-term planning issues. Rational technology assessment is dedicated to this challenge, balancing long-term orientation and short-term flexibility requirements. It should enable the assessment of „visions" with respect to both their relation to long-term stability, rationality and ethical implications and to their relation to short-ranged acceptance and flexibility requirements. The path of rationality in this field is shown to be, as in most cases, the pragmatic midway and the result of a careful weighing process. Rational technology policy consists of contextual judgements relating the issues of the singular decision situation to underlying (culturally dependent) normative rationality standards embedded in society.

Subsequently, a case is discussed in which rational technology assessment may offer a fruitful way to shape developments already in an early stage. Michael Decker writes about the use of robotics in health care. "Autonomous robots are developed to replace human beings" could be a provocative statement to call for Technology Assessment (TA) on these applications. Prototypes in operation in the laboratories of robotics today are in early stages of development, but not in the focus of societal interest up to now. It is widely agreed that TA of new technologies should start in these early stages of development. It is also well known that there is a dilemma between "early stages of development" and the study of consequences of technological advance for which a detailed knowledge about new technologies is necessary. In this paper the proposal is made that the development of an interdisciplinary vision can be the basis for technology assessment in the sense of "vision assessment". Therefore an interdisciplinary expert group should be called together in which the perspective of the developers of a new technology can be combined with perspectives of other scientific fields found to be relevant. Experts should be selected according to two different criteria. Firstly the relevant sciences must be determined by structuring the field of interest. Secondly the experts must be interested in interdisciplinary work and should agree to a set of rules upon which interdisciplinary discussions are to be based. In this discussion process, each scientist has to convince the experts of other disciplines of his perspective by argumentation and at the same time the underlying assumptions of the individual scientific perspectives have to be revealed in order to combine all perspectives and thus to reach a balanced interdisciplinary vision. This balanced vision can be used both to initiate a participation process, in which laypersons should add their perspectives, and also to influence the political decision process.

In this contribution the establishment of such an expert group is outlined for the field of robotics in health care. A structuring grid is presented, which can be used as a basis for discussion and for the decision as to which sciences should participate. Finally a proposal is made on how the expert group should work in order to prepare a sensible contribution to Rational Technology Assessment in this context. The chapter sheds light on the complexities and opportunities of that approach in an area in which developments are on early stages – that is, where Rational TA is both most adequate and most promising. Due to the fact that the chapter is concerned with a future TA-project some aspects are presented, which appear less clear in an *ex post* analysis.

Finally, in a joint undertaking by all authors, lessons from their various contributions are drawn. First, it is argued that the suspicion that visions in many sectors reflect a certain dominance of the assumptions of High Modernity has, by and large, been correct; and that we should hardly be surprised about that finding, given that structuration processes of various kinds, of which we have met several in the case studies.

Subsequently, the issue of long term planning, attempting to bring together the considerations brought forward by Grin and by Grunwald as well as lessons and illustrations from the other chapters are dealt with. This section is intended to be a contribution to some necessary reflection within the TA community on the implications for TA of the debates on long term pianning (for instance with regard to realising sustainable development) and on modernity/post-modernity.

In a more practice-oriented part there are some methods and techniques presented from the various chapters to uncover the assumptions underlying existing visions, as well as methods and techniques to assess these visions critically and constructively. The intention has been to present a first outline of a repertoire of tools that may help TA to become a valuable assets in the societal debate about the future.

I   Introduction

Vision Assessment to Support Shaping 21st Century Society? Technology Assessment as a Tool for Political Judgement

John Grin[1]

1 Shaping the Future?!?

This book is about ways to think, as well as on ways to think about thinking, about the future. Specifically, we will focus on the role of technology in the first half of the 21st century, as we are supposedly entering the 'post-industrial,' 'trans-national,' 'post-modern' (and so on) society. And we will ask how we can critically assess and construct visions on the ways in which technology and social problems are going to relate to each other.

That undertaking relates to two major debates that have been going on the last quarter of the 20th century. The first one, obviously, concerns the nature of the society that we conceive for ourselves during the 21st century. It is concerned with questions like: What should we do in the 'age of side effects'? (Beck 1992; 1997) What sense does it make to say, as e.g. Peter Drucker (1993) and the Toffler couple (1980; 1994) are doing, that money and violence are increasingly replaced as instruments of power by information in its various expressions? What future is there for the sovereign nation state and related concepts and institutions; and what does it mean to talk about 'inter-national order' and 'the international community'? What are the sense and non-sense of proposals to go for a factor four, ten or even twenty reduction of the environmental burden of the ways of production and consumption used by the world population to fulfil human needs? (Vergragt and Jansen 1993; von Weizsäcker et al. 1997) Et cetera.

But in thinking about the societies we wish and do not wish for ourselves, it is unavoidable that we, secondly, run into the debate on the 'malleability' of society. *Can* we actually, somehow, shape our society? Societal processes, including the development and use of technology are, essentially, institutionally embedded multi-actor processes, without any single actor being able to determine what is to happen. Even governments have run into major limitations. First, they lack the

---

[1] The author is indebted to professor Egbert Boeker as well as to the other authors in this volume, especially Armin Grunwald and Rob Reuzel, for their comments on an earlier version of this introduction.

capabilities required to centrally and unilaterally steer the actions and interactions of (transnational) non-governmental and business organisations. And, second, the legitimacy of governmental action is no longer ensured simply by a combination of an authoritative knowledge base and democratic elections. But if we thus cannot *shape* our societies, what sense is there in thinking about what society we *wish*? Or is it possible may be, not to steer, but to influence our future from some normative perspective? It is fair to say that such disciplines as policy science and technology dynamics have, the past few decades, been dominated by dilemmas that relate to the tension between our wish to shape our own future - the good old rationale of democracy - and the actual problems in doing so. This book is not free of such dilemmas, but, on the contrary, it is an attempt to explore ways to deal with them more or less fruitfully. The chapters of Mambrey and, especially, Grunwald explicitly deal with these issues.

How deep-cutting these two debates are, especially when they are interrelated, can be exemplified by two recent Dutch books on utopism. Crombag and van Dun (1997) discuss the "utopian temptation". They insist that utopian authors "suggest to us an in principle completely controllable and malleable world that meets the specification of a model, based on the most recent scientific insights" (1997, p 11). They hold that social problems are thus reduced to scientific questions, and that designing a world on the basis of the scientific answers to these problems implies that human diversity is being denied. Thus utopias are ideals being forced upon reality, and thus not truly human ideals.

The author of the other book, Hans Achterhuis (1998) departs from a more sympathetic viewpoint, starting his work with the confession "[d]eep fascination and intense repulsion: these carried and inspired my quest into Utopia" (1998, p 11). Yet, within just over one page he informs the reader that having read Ernst Callenbach's *Ecotopia* (1975) with the interest of someone engaged since long in the environmental issue, ends up with wondering why he would never wish to live in Ecotopia.

Thus, in both books utopism is discussed from the perspective of the question: to what extent do utopia's presume that *all* comply with the ideal society dreamt up by *some*? Even after Sir Karl Popper's (1963) fierce and fundamental criticism of utopism, the conclusions are de-masking in many ways. But simply taking this as just an ultimate verdict over utopian views misses the drama. While utopias usually have unmistakably authoritarian characteristics, the fact is that most utopias originate from deep and legitimate concern about the world as it is, and an equally deep and legitimate desire for a radically better world.

Can we do better? Can we construct futures which are attractive to all of us and which can, moreover, be realised? Let us start with the latter question, on feasibility, and see whether we can subsequently say something on the acceptability issues. As I have already indicated, in technology dynamics much attention has been paid to the challenge of shaping technology development, in spite of the inherent uncertainty on the social impacts of technologies and the long lead times involved. It is not in the least place against this background that there has been much emphasis within the field on the ways in which technology and its social context are being shaped along with each other, on the basis of a set of common or at least mutually congruent expectations. These expectations concerning technology, its social context and the relations between the two, are projections for the future, rooted in actor's assessments of past experiences. Both evaluations of the past and expectations for the future reflect the values, worldviews and deep preferences of those who hold them. In other words, they represent the bounded rationality of their 'possessor'. Thus expectations are not arbitrary, but remain within a particular 'horizon of expectation,' to use the phenomenological term. These horizons delimit the range of people's *'futuribles'* (attainable futures; *cf.* de Jouvenel 1963). Such *futuribles* we will call 'visions' throughout this book, following the designation used in some recent path-breaking work in Germany (Mambrey et al. 1995; Dierkes et al. 1996).

For a more thorough definitional and theoretical discussion I am happy to refer the reader to the chapter by one of those pioneers, Peter Mambrey. Let me here just mention that visions have two major features: they are a mental image of an attainable future shared by a collection of actors; and they guide the actions of and interactions between those actors. The German original term may be more clarifying here than its English translation: in German, a vision is called a *Leitbild*, a 'guiding image.' This term entails a direct reference to the musicological term *Leitmotiv*: a universally recognised musical theme that creates associations among a variety of listeners to some extra-musical phenomenon known to each of them, such as the loss of a beloved or erotic experience. Similarly, a technological *Leitbild* or vision is a recognisable future view that creates 'congruent' (*cf.* Grin and van de Graaf 1996a) actions among a variety of actors involved in its realisation.

One might say that the notion of visions suggests that the feasibility issue be solved by stressing the empirical observation that people *are* being guided by shared visions as the basis for collective action. That is, we consider visions as „common language" (Habermas) that guide collective actions. To be sure, we acknowledge that in the course of High Modernity particular visions have become rather dominant as determinants of our future. Yet, the fact that we do *construct* our visions and the meanings implied by them is to say our societal systems are completely self-referential (Luhmann): by critically reflecting upon our visions, we may be able to shape them and this to influence societal development. **In other words: the question is *not* whether it is possible to shape the future according to some shared vision, but rather whether it is possible to shape the visions that *are* guiding us into ones that we like better. Another question is how to shape visions without running into the acceptability problems that we just indicated for existing utopia's.**

These are obviously complex questions. Especially when we are considering areas, such as promoting truly sustainable development, where a prolonged process of change is desired, pertinent normative and epistemological questions arise. Later in this book, Armin Grunwald deals with such issues in a rather fundamental and thoughtful way. Specifically, he attempts to shed new light on the question how to reconcile long-term planning with short-term flexibility and public acceptability; and he is looking for some form of normative grounding of the answers. For instance, how to realise a factor 10 or 20 reduction in the environmental burden associated with fulfilling needs such as food or housing, given that we generally know so little about the future? And how to prevent that attempts to do so degenerate into utopias that have similarly authoritarian implications as Achterhuis as well as Crombag and van Dun have revealed earlier utopias?

Against this background, we can now more accurately formulate the central question for the pages that follow: *To what extent and in what ways can TA contribute to shaping the future through critical and constructive assessment of existing visions and the assumptions underlying them?*

'Constructive' has a dual meaning here: it indicates the primary intention (bringing about a better future rather than merely cynically breaking down existing dreams) *and* it indicates the desire to have TA contribute to *constructing* visions that guide the ways in which actors construct their actions. 'Critical' of course indicates that we are looking for TA approaches that help to reveal and evaluate the assumptions underlying visions and that take little for granted. This immediately raises, of course, the question: 'criticism from what perspective'? And, if we had not discovered it already, this question directly confronts us with the analogous one on the construction of alternative visions: 'constructions from what perspective?' Let us explore these questions in the next section, departing from a recent, thought provoking comment on some visions concerning the world food problematic.

2   Existing Visions, Modernity and Criticism

On December 18, 1998, Professor Louise O. Fresco, director of research, public relations and natural resources of the United Nations Food and Agriculture Organisation (FAO, Rome) gave the 27th Huizinga lecture (Fresco 1998). In the trail of both Johan Huizinga and earlier Huizinga lecturers, her talk was wide-ranging, future oriented, focusing on the threshold between past and future and both provocative and thoughtful. In one word: she presented a *vision*.

At first sight it may seem more accurate to say that she *discussed* visions formulated by *others* on the future of global food production. The first vision she illustrates with the portrayal of western food production in a picture '*Extraterrestrial mutants arrive with UN food aid*' by the Cuban painter Julio Breff Guilarte. Two female giants, crossings between women and chickens, have just laid some dinousaur-like eggs and look down on a desoriented, apparently powerless, farmer with an arrogant smile on their faces. The picture thus expresses the threat of the technological invasion embodied by contemporary western 'high-tech' agriculture. It represents a fear, designated 'Shadow Thinking' by Fresco, that at the end of the 20th century has become widespread. That fear reflects a deep pessimism concerning technological progress. It finds a variety of expressions:

- neo-Malthusian fears that we are about to face serious food shortages;
- ecological fears on the damage for our planet as a consequence of feeding a strongly increasing number of mouths,
- fears that mankind is poisoning itself.

Shadow Thinkers put the blame for these developments on the shoulders of the powerful troops of science and technology; its strongest contemporary expression is found in the resistance against biotechnology, especially genetically modified organisms.

Fresco, reminding us of Huizinga's notion that, historically, 'doom thinkers' have always tended to idealise the 'purity' of 'the' past, then presents the step-brother of Shadow-Thinking: Light Seeking. Light Seekers emphasise the romantic appeal of fairness, serenity and simplicity of nature, body and mind. They can be found amongst Montignac-adherents, New Age adherents, 'bio-fooders' and so on. Their thinking has not so much anti-technological, but rather anti-rational tendencies. Simultaneously, Light Seekers launch alternative, theoretical concepts around the absolutist notion of 'naturalness', and attempt to live alternative ways of life around these concepts.

Louise Fresco confesses that she does not in the least share the technological pessimism of the Shadow Thinkers, but that she nevertheless sees an important role for them: they warn the rest of us timely for dangers initially overlooked by the scientific and technological elite. Thus their gloomy predictions have a value, not as self-fulfilling prophecies, but as what she calls self-denying prophecies. Similarly, while she holds Light Seekers' views for simplistic and anti-rational, she points to the sense of undermining the self-evidence of existing consumption patterns.

But if the two visions of Shadow Thinking and Light Seeking can both be criticised and have something to contribute, obvious questions are, of course, *from which perspective* they can be criticised, and *to what* they may present a valuable contribution. One would, therefore, expect at least a third vision to be present in Fresco's lecture. Indeed there is: implicitly throughout, and explicitly at the beginning, where she presents a range of data to indicate the qualitative and quantitative improvements in food production over (especially the second half of) the 20th century. Her own vision also becomes explicit at the end, where she emphasises the capabilities of the scientific-technological complex to meet the challenge of responsibly feeding the growing world population, using its adaptability and openness to take into account the more useful insights from Shadow Thinkers and Light Seekers.

The relation between the three visions is unambiguously clear. As Fresco sees it, the third one should be allowed dominance. The other two may be useful supplements, through providing warnings (the Shadow Thinkers) or in contributing to bringing about a necessary change in consumption patterns as an important complement to more efficient and qualitatively better production technologies (the Light Seekers). But it are these technologies that are central in her vision. Having mentioned the contributions of Shadow Thinking and Light Seeking, she immediately adds the stipulation that these do *not* amount to sources "to develop alternative foundations for the sciences" (1998, p 53). Rather, what we need is a "new contract" (1998, p 51) between science and society, a contract that ensures the continued legitimacy of science and technology. This contract should take away anti-technological fears by taking into account stimulating insights from Shadow

Thinkers and Light Seekers. And it should encourage science to play its role in society enthusiastically, outlining the contours of new political thinking.

Louise Fresco ends her lecture with the assertion that it are such visions that feed us, keep us moving. The fact that the visions for tomorrow present their shadows today, should not make us stand still and idealise the past. In a final Huizinga-quote: "this we know - there is no general return. There merely is advance, even if we are being dizzled by unknown depths and distances" (Huizinga 1935, p 12, as quoted by Fresco 1998, p 54). In sum, the third vision reflects the well-known faith in scientific and technological progress in order to solve societal problems. It is the same faith that also underlies many of the utopias considered so critically in the already cited books by Crombag and van Dun (1997) and by Achterhuis (1998), and which puts technology at the lead in creating a new world, putting humans in the follow-mode. An appropriate designation for this typically Modern approach may be the 'True Belief' (although Professor Fresco might prefer some less distanced designation, like 'Rational Thinking'). Thus seen, a much deeper appreciation emerges of what really is at stake in Fresco's discussion of three alternative visions on the future food system - a discussion which she herself is tempted to see as a "scale model" (1998, p 51) of the relation between technology and social problems more generally. The confrontation between the three visions is nothing less than the collision between core ideas in the debate on Modernity.

But if this is true, have we then answered the questions at the end of the previous section? Is it valid to say that the particular mix of visions proposed by Fresco is an example of how a new vision can be constructed out of existing ones – a vision that is more widely acceptable and more robust against side effects than the True Believers' vision *per se*? In section 4, I will argue that the answers to the latter question must be no. We need a more fundamental kind of assessment of existing visions to have a basis for constructing new visions for the 21st century. Only then we will be able to benefit from what we have learnt from the experiences during that 'Age of Extremes,' this 'Short Twentieth Century' (Hobsbawm 1994) that has brought us more dystopian novels than utopian ones, as Boeker (1975, p 118) has noted. To appreciate that need for fundamentally different approaches to vision assessment we must, first of all, appreciate the nature of the knowledge claims that underlie Fresco's assessments of her three visions. That is the subject of the *intermezzo* provided in the next section.

3   Knowledge in the 'Age of And'

In order to appreciate how the visions distinguished by Louise Fresco have more in common than may occur at first sight, let us relate them to three core concepts in Ulrich Beck's (1997) theory of reflexive modernisation. The Shadow Thinkers are those who emphasise risks and side effects, in 'risk society' during 'the Age of Side Effects'. The True Believers represent the view of Simple Modernisation: linear advance, instrumental rationality, and ratio, certitude and understanding as the basis of action. Light Seekers, finally, represent counter-modernity, which emphasises emotions like love and fear as a basis for thought and action and which chants the past. Beck emphasises that modernity and counter-modernity have always gone together, in a dialectical relationship.[2] The co-existence of modernity and counter-modernity has given rise to constructed certitudes such as naturalness or gender roles. While they are, substantively, rather a-modern, these constructed certitudes, by acting as Archimedal points, help to legitimise counter-modern views in the face of modern adversaries.

This draws our attention to a fundamental similarity between True Believers and Light Seekers: both defend truth claims in the form of universal certitudes. Shadow Thinkers, especially as Fresco describes them, in fact hold similar beliefs, the main difference being that their generic truth is more pessimistic in nature. Fresco merely privileges one of these beliefs. While she acknowledges some role for Shadow Thinking and Light Seeking, she does so merely to the extent that they appear to make sense from the perspective of the True Belief. But how can one justify such a choice? Is there any objective standard that informs us that what Fresco holds to be the True belief indeed represents the ultimate truth, against which all other knowledge claims can be measured?

Of course, the conviction that this can be done is typically Cartesian, as is the idea that true knowledge sees the world as composed of systems of objects, governed by laws that lead to stability, order and hierarchy. The exemplary science corresponding to these principles is classical physics, most specifically Newtonian mechanics with its laws of motion; another example is Euclidean geometry with its logical structure of axioms and laws. Stephen Toulmin (1991, pp 180-181) has reminded us that Descartes also postulated that the science he propagated contained its own standards of truth: a Benevolent God had planted the basic concept of Newtonian mechanics and Euclidean geometry in the minds of people from all cultures and epochs. In other words, Cartesian philosophy asserts that there is an objective 'scratch line' from which one can define true knowledge.

In this light, Fresco appears a true follower of Descartes when she appears to hold that visions constituted on modern agricultural science, based on molecular biology and similar fields, represents the standards against which all other visions can be assessed. However, this assertion neglects the idea that, as Stephen Toulmin immediately adds to his discussion of Cartesian truth claims, that the 'burden of proof has now shifted': there is no scratch line. The fact that Toulmin hardly substantiates this claim does not, by itself, make it implausible. Richard Bernstein (1983, especially pp 126-144), in his monumental and enlightening discussion of the road from positivism to hermeneutics, has provided a convincing discussion on precisely this point.

Perhaps the most clarifying example Bernstein (1983, pp 131-132) mentions is a famous autobiographical anecdote from Thomas Kuhn, who in the summer of 1947 finally managed to make sense of Aristotle's physics. Until then, he had

---

[2] In passing, it is interesting to note that early examples can be found in 17th century utopia's such as Jean Valentin Andreae's (1619) *Christianopolis,* that often included magical and alchemist elements in addition to what many now regard as proper science. Today, we find a similar mix in ideas like Theodor Roszak's (1992) 'Ecopsychology' or James Lovelock's Gaia thesis (1979; 1986).

wondered how such an "acute and naturalistic observer" could have said so many absurd things about motion? (...) And, above all, why had his views been taken so seriously for so long a time by so many of his successors?" But, he continues, "one memorable (and very hot) summer day those perplexities suddenly vanished" when he discovered an alternative way of reading Aristotle's texts. At the core of these ancient ideas were not so much moving material bodies, but change-of-quality in general, including both the fall of a stone and the growth of a child to adulthood. Once Kuhn had grasped that point, he was able to appreciate Aristotle's views on motion as rational in their own right.

Bernstein then goes on to argue that Kuhn's discovery is at the common core of the insights provided by a range of other authors that have contributed to 20th century hermeneutics: Thomas Kuhn, Imre Lakatos, Peter Winch, Clifford Geertz and Hans-Georg Gadamer. Bernstein's synthesis of these authors' views contains the idea that an analyst's task should be the attempt to "find the resources in our language and experience to enable us to understand [these] initially alien phenomena without imposing blind or distortive prejudices on them" (Bernstein 1992, pp 141-142) This requires that the analyst does not subsume objects under scrutiny under given standards, generic 'prejudices' or, as Bernstein prefers, prejudgements that contain truth. Rather (following especially Gadamer here) the task should be understanding or judgement: discovering the meaning of the object by looking at it from one's 'prejudices'; simultaneously improving one's appreciation of these prejudices by seeing what they bring about in applying them to that object. Because prejudices always play a role, there is no *a priori* distinction between scientific and other knowledge traditions. Neither can any knowledge system claim any *a priori* right to provide the standards against which all others can be measured. *If* we can ground our judgements, it is not in terms of a universal standard, but only in terms of standards provided by our own culture (compare Grunwald, this volume).

But does this leave us with total relativism? Are we now forced to accept any knowledge claim? Does this totally deny the sense of the science as a sound basis for thinking about the future? The answer to these and similar questions is 'no, because...' The 'because' is so difficult to appreciate since it relies on the recognition that we need to go beyond what Bernstein calls the 'Cartesian Anxiety,' the fear that we are lost without a fixed, Archimedal point. Beyond that Anxiety, however, is the perspective of finding truth - not in terms of universally valid theoretical claims, but in terms of contextual wisdom (compare Janich 1996)

Beck (1997, pp 1-3) catches this point by referring to Wassily Kandinsky, who as early as 1927 wrote about the transition from 'the age of *Either-Or*' to 'the age of *And.*' It is no longer advisable to state the problems as one between either recognising the authority of science, or be lost into relativism. Rather, our current age should be one characterised by the recognition that wisdom cannot be defined on the level of generic truth, but rather on sensible judgements to inspire action in particular contexts. If we are to save ourselves from the condition of current risk societies by guiding ourselves into a more 'reflexive modernisation,' we need to re-discover judgement in the trail of Aristotle.

Here, Beck follows Stephen Toulmin (1991) in his re-appreciation of Michel de Montaigne. Toulmin, on good grounds, argues that Montaigne may well be seen as the second great thinker, next to René Descartes, of the two centuries following the Middle Ages, especially the bloody and chaotic fourteenth century (cf. Barbara Tuchman's (1979) vivid historiography). It has become a widely accepted thesis that the Cartesian quest for certainty was a response to this state of affairs. Toulmin attempts to add plausibility to that thesis by reporting about his search for evidence that the young Descartes was greatly touched by the feeling of chaos following the murder on Henri IV of France. More importantly, however, he draws attention to Montaigne's alternative way of responding to the chaos.

It was precisely Montaigne's type of non-doctrinal thinking that is reflected in the Edict of Nantes drafted by Henri IV, and emphasising tolerance as a means to prevent religious disputes leading to conflict and chaos. Toulmin (1991, p 29) stresses that we should understand Montaigne's 'skepticism' *not* in a Cartesian way, that is as a form of

> "destructive nay-saying: the skeptic denies the things that other philosophers assert. (...) [Skepticists] no more wished to deny general philosophical theses than they wished to assert them. (...) [they] saw philosophical questions as reaching beyond the scope of experience in an indefensible way. Faced with abstract, universal, timeless theoretical propositions, they saw no sufficient basis in experience, either for asserting. or for denying them" *(emphases in the original - JG)*.

To give one specific, and in this book central, example: skepticist will not sign to the conviction that technology *per se* will be able to solve social problems. Rather they will judge, case by case, whether this is possible, given the particular social and political conditions (see e.g. Toulmin 1991, pp 180-184; Beck 1997, pp 161-163).

While there is undoubtedly reason to regard Toulmin's classification of Montaigne-humanism-skepticims as 16th, and Descartes-science-rationalism as 17th century thinking as historically too simplistic, this does not at all undermine his thesis that *two* streams of thought attended the cradle of Modernity. In addition to what became to be known as *the* Modern view, there also were the skepticists. Thus, the "new constellation" of modern and post-modern ideas that Bernstein (1992) perceives at the threshold between the 20th and the 21st century may be seen as a reflection of a similar constellation existing in 16th and 17th century. Of course,, in a sense, this constellation has persisted since. Yet it has taken until well into the 20th century before the Montaigne-stream started to receive a degree of appreciation that at least resembles that enjoyed by the, hitherto hegemonian, Cartesian stream.

That this shift may, under circumstances, concern a wide variety of dimensions of science, technology and society becomes clear in the chapters that follow. Mambrey shows that technological expectations and assumptions on organisational functioning go hand in hand in existing technological visions on the use of computer technology in offices. Reuzel and van der Wilt show how the centrality of established medical knowledge, the control of medical experts on medical technology development and the nature of governmental policy on medical treatment packages form a coherent, self-reinforcing whole. Similarly, my own contribution on visions guiding air power development show that in order to make different choices concerning technology development we need also different approaches to public legitimacy and the role of nation-states in international (transnational?) security.

Toulmin (1990, pp 175-201) deserves the credit for having drawn our attention to the fact that, while most established scientific disciplines have long represented Cartesian knowledge ideals, this is now over. Increasingly, scientific knowledge looks for truth *not* on the level of abstract, general truth, but rather on the level of being able to describe, contextually, physical processes. One example he mentions (1990, p 181) is that of molecular biology. Far from being just another form of mechanistic science, it appeared to be most fruitful when applied to understand biochemical processes as they occur in their 'microhabitat' within the body. Rather than being pre-occupied with universality and uniformity, such science emphasises understanding diversity and adaptability. Similarly, to add another example, we may point to developments during the last two decades in economical sciences. In that discipline, classical economy with its emphasis on market equilibria and actors that all have the same, profit-maximising, rationality has been challenged by so-called institutional or evolutionary economics. As two of the latter approach's pioneers remark, "it is no caricature of orthodoxy [neo-calssical economic theory - JG] to remark that continued reliance on equilibrium analysis, even in its more flexible forms, still leaves the discipline largely blind to phenomena associated with historical change" (Nelson and Winter 1982, p 8). With the evolutionary they develop, they wish to contribute to remedying that omission.

Such examples illustrate that even *within* the sciences, there has been renewed appreciation of contextual judgement, doing justice to adaptability and variation. Although I would haste to add that generic knowledge, in the form of universal theoretical propositions, definitely makes sense in various fields and circumstances, this approach has now lost its sacrosanctity. Yet, to fully appreciate the type of judgement that lies at the basis of reflexive modernisation, we need go beyond the way in which knowledge is constructed *within* disciplines and other forms of social organisation. That is, we need to explore the possibility that wisdom may be contextually discovered *between* a variety of knowledge systems, more or less Cartesian in nature. Let us discuss this in the next section, focusing on the issue of how to construct 'wise' visions.

4   Constructing Wise Visions in the 'Age of And'

It is clear from the above that in the hermeneutic view, contrary to the established positivist view, traditions that shape our prejudices are not something to be excluded but resources to be explored through the process of understanding. This is possible because, when we apply our prejudices to specific objects, we do so in order to define our courses of action. In that sense, we are - conversely - able to appreciate our traditions and past experiences by assessing what they make us do in our context. Traditions thus shape our *horizons*. In the terminology of Hans-Georg Gadamer, our task is to strive for a 'fusion on horizons.' This does *not* merely mean to add to our own horizon elements from other ones. It includes a reflection on our own horizon, its prejudices and history as much as a reflection on these other horizons, in their own right. We thus enlarge and enrich our horizon, recognising that a horizon is not fixed and limiting, but rather something "into which we move and that moves with us." (Gadamer 1975, p 271, quoted by Bernstein (1983, p 143). In doing so, we will also be able to more fully appropriate our own prejudices. This idea of fusing horizons represents a truly skeptical approach to synthesising visions so as to take lessons from the past into account when thinking about the future. Let me emphasise here that I am *not* saying: a skeptical vision. There is a very fundamental reason for this. If we can talk about a skeptical vision in any way, it is definitely not *generically* - the Skepticists' central claim is that no generic wisdom is possible. We can only understand a specific situation *in its own right*; and we can base sensible action only on such situational understanding.

That is, by their nature, we can only identify the substance of skepticist visions contextually, through some form of practical reasoning. While this may appear unsatisfactory to many of us who look for generically valid solutions, it may also have important advantages. We will explore this in the remainder of this section, which is devoted to discussing some of the core implications.

4.1
The Need for Contextual Assessment

First, the analysis should be performed contextually. Even if the research problem would be formulated relatively generically it would have to be answered through some form of contextual judgement. For instance, assume that the commissioner of the TA is interested in the future of the old vision that information technology would create a 'paperless office.' She might formulate the research problem rather generically as: *Given that e-mail and internet use is now widespread, are we eventually able to realise the paperless office?* Then a proper way to answer it would be to perform case studies on several types of offices, selected in a way that would make it possible to explore whether a generic answer can be given at all. The result might be a generic answer to the generic question posed, or a sound argument why this is not possible, or a typology of cases and answers per type.

What this means, for the methodology of TA as vision assessment, is a question that is more urgent than its answer is obvious. The reader is cordially invited to join the quest we are engaging in throughout the following chapters to explore that issue. The conclusions we draw in our joint final chapter are intended, of course, as an attempt to shed some light on these issues; but, especially, as a further invitation. It is probably fair to say that such reflection amongst TA researchers and other policy analysts is highly needed if their professional fields are to maintain their relevance under 'the post-modern condition.'

One important implication must be mentioned here, however. We are facing the possibility that *visions themselves need be seen as a contextual concept*. In fact, in many of the cases discussed in the German literature on *Leitbilder*, visions concern the network of those involved in, e.g., the development and early use of a particular technology, such as the typewriter or the Diesel engine (Mambrey *et al.* 1995; Dierkes *et al.* 1996). Thus rather than formulating and peddling generic visions, it may be better to focus on the construction of contextual visions. Contexts here may be understood here both 'geographically' and 'socio-cognitively'. To give a - fictive - example of geographically different visions: biological cultivation appears most appropriate for potatoes in the South of the Netherlands, while, simultaneously, more high tech cultivation may be more adequate for potatoes in the East of Germany or for carrots in the South of the Netherlands. At least, such conclusions may follow from an assessment that takes into account the races of the groceries involved that can be grown on the soil of the respective regions. This means that, from the plurality of views characterising advanced, democratic societies, some may be favoured most at some instances and others in other places. This may mitigate some of the concerns mentioned in section 1 against the authoritarian nature of utopian thinking.

To give an example of 'social-cognitive' context from the same field: it may be possible to establish two different market-chains for tomatoes that each by itself as well as together are better in line with the idea of sustainable development. One chain would be that of growing, processing and selling biological tomatoes for a sub-set of consumers who do not care about the fact that its availability will be spread over a smaller part of the year, as well as for food producers which use it as an ingredient. The other would serve those consumers who prefer wider availability. In other words, this type of differentiation gives room for plurality in future visions as just another way to meet the criticism that has been raised against utopian thinking.

Of course, in those cases in which we are used to centrally taken, generic decisions, such contextuality may pose non-trivial difficulties. This appears to be the case, for instance, in the cases of medical technology (Reuzel and van der Wilt in this volume) and military technology (Grin in this volume). We will reflect in the final chapter on the problems and opportunities we have identified in these respects.

4.2
TA to Support Political Judgement

Second, TA as vision assessment that leads to a fusion of horizons should contribute to practical deliberation. This statement has considerably more depth than may appear at first sight. May be the best starting point for exploring its significance is to stress that practical deliberation *cannot* be sensibly done by a single analyst in the person of a skepticist *pur sang*. Not only, first, do individuals without prejudices not exist. Even John Laursen readily admits this when he points out that Montaigne, whose "tranquillity" and non-dogmatic thinking he discusses with admiration, "was not amoral" but, for instance, departed from a "hatred for cruelty" in many of the positions he took (Laursen 1992, p 100).

Second, once seriously attempted, an approach of completely detached criticism leads to major normative difficulties. Bernstein (1992), in a final reflection on a fascinating set of comments on the most prominent among post-modern thinkers, notes that

> "[i]t is striking that Foucault, Derrida and Rorty – after their abstract negations – turn in their later writings to struggling with ethical-political questions. And this is not (...) simply a 'contingent' fact about the 'development' of their thinking, but rather it is an indication of their efforts to face up to the consequences of their *own* thinking" (1992, p 311).

To be sure, the main merit of the post-modernists is that they have shaken our confidence in the rational grounding of critique. Contrary to Hegel, Marx and others they hold that critique cannot be properly grounded. Yet, he adds, leaving it with that unavoidably leads to cynicism and thus ultimately to cruelty, as Richard Rorty's attempts to base political positions on 'irony' indicate. Denying the possibility to ground our critique, however, is not necessarily the same as discarding the possibility to the practice of critique altogether. We just need to recognise that

> "there cannot be any critique without some form of affirmation, that we cannot avoid asking the question 'critique in the name of what?' And we need to be prepared to defend out affirmations and standards of critique when they are challenged" (1992, p 318).

Bernstein is undoubtedly aware that he thus makes an important move: to go beyond modern *epistemology* he turns into the realm of the *political*. Even stronger: once we truly recognise what it means to leave the modern Either/Or, we see that "exorcis[ing] the quest for certainty and certitude (...) we [need to] engage in critique as second person *participants* and not as third persons neutral observers" (Bernstein 1992, p 319). Yet precisely at this point we need to be more accurate. In his standard work on political judgement, Richard Beiner (1983, pp 102-128) has pointed out that a basic characteristic of political judgement is that it involves both the activity of what he calls the distanced spectator and that of the involved actor. He holds that

> "[t]o forfeit the standpoint of the spectator is to sacrifice dignity, which arises from distancing oneself from the particularity of natural purposes. To forfeit the standpoint of the agent is to sacrifice wisdom, which comes from long and rich pursuit of human ends"(1983, p 109).

The core of Beiner's argument, and probably also the cause for him being more nuanced than Bernstein, is basically empirical. He holds that these two types of reflective judgement cannot be separated, but are dialectically inter-related and, empirically, presuppose each other in terms of their conditions. He illustrates his point with the biographies of two ancient personifications of either type for judgement. Pericles, the prototype of sound practical judgement-in-action, was capable of detached judgement, and it was that capability which was of great guidance in deciding on political action. And Thucydides was able of sound, detached judgement just because he knew the Athenian political life from which he had been exiled so well. It is interesting to note that Laursen (1992, pp 94-96) reminds us that the latter was also true for Michèl Montaigne, who was engaged with politics during his lifetime and who fought on the catholic side during the French civil wars.

Based on this appreciation of practical deliberation as essentially a form of *political judgement*, we can now explore the methodological implications of our demand that TA should support the 'fusion of horizons,' that is: constructing visions through political judgement.

4.1.1
Visions and their Basic Assumptions

One entrance into these issues is to wonder what it means for TA to support a process in which people, first, uncover the assumptions underlying their own and each other's visions, and, second, discuss visions in these terms. The work of Frank Fischer (1980; 1995), in the hermeneutic tradition in the policy sciences, has provided some concepts that may be helpful in order to identify and order such assumptions. In describing the rationality informing a particular actor's policy as a plan-for-action, Fischer distinguishes between first order notions, which are specific for the context of that action, and the more generic second order convictions underlying them. First order notions include two layers: assessments of the means to achieve given objectives (solution assessments), and problem definitions that contextually vindicate these objectives. Second order notions comprise on the one hand the world views and value systems and, on the other hand, the preferred social order.

First order notions can be identified relatively easily. One may directly inquire into solution assessments (what are, for that actor, the costs, effects, and side effects of a particular solution?). Similarly, one can directly inquire into problem definitions (what does the actor consider to be the problem or 'challenge,' what criteria does she use to judge the situation and how does she weigh them?). Although one may also be able to inquire directly into second order convictions, this may be more difficult, if only because such underlying assumptions may be 'hidden' for the actor herself. Elsewhere (Grin *et al.* 1997, pp 32-33; 60-65) we have argued that one way to tackle this problem is to complement direct inquiry with asking 'why' questions. By asking why somebody considers a particular aspect an advantage, or why he gives more weight to the claimed advantages than to the admitted disadvantages, his problem definition may be revealed, including the importance attached to its respective dimensions. Similarly, by asking why the problem is defined in a particular way, one may identify the relevant elements of the world views and value systems underlying them. And asking why these values and causalities are considered important is likely to yield an answer in terms of "because they guide actions so that we get closer to the world I prefer." Later chapters in this book will discuss those key issues in technological visions that may be particularly suitable for identifying crucial assumptions; we will collect the lessons in section 7.3.

The table below is intended to illustrate these concepts as well as to offer some clues to appreciate the nature of the assumptions in specific visions (in the chapters which follow as well as in societal debate). It offers some important features of the True Belief, Shadow Thinking and Light Seeking that may underlie technological visions. It is just based on some reflection on Fresco's lecture from the perspective of Beck, Bernstein, Toulmin and others' contributions to the ongoing debate on Modernity. The table sketches some important elements of what Toulmin has called the *cosmopolis*: a coherent set of views on nature (the *cosmos*) and the polity *(polis)*. Of the four levels distinguished by Fischer, it contains only three. It yields ultimate preferences, world views and values systems, as well as dominant elements of problem definitions for these three types of thinking. While the three perspectives share, as we have seen, the typically Modern notion of certitude, they differ in many other respects, including in the nature of the certitude on which they focus.

The table may be used, first, to appreciate the nature of some important underlying assumptions in specific cases. Thus, in order to identify crucial assumptions, one may think back and forth between the table and 'why-questions' concerning the key issues in technological visions, to be identified in later chapters. Second, the table may be used in order to assess a particular vision *A* in terms of the basic assumptions underlying another one *B*. To do so, one should assess key elements of vision *A* in terms of the criteria (problem dimensions) considered crucial by adherents of vision *B*.

Table 1: Three ideal typical belief systems on the *cosmopolis* The table shows the author's reflection on a lecture by Louise Fresco (1998) on visions in the area of agriculture and world food production that serves as a 'scale model' for other technological visions.

Item	True Belief	Shadow Thinking	Light Seeking
Preferred social order			
View on the polis	Leviathan: State to ensure social stability through central government, based on certain knowledge.	'The intelligence of democracy' (Lindblom 1968): checks and balances, employing counter-expertise to guide governance.	The republic: responsible citizens as a basis.
Role of citizens	Loyal citizens.	Citizens check for balances, and are loyal if this is ensured.	Should develop and act on the basis of a collective moral that promises good for the whole.
(To be continued next page)			

World views and value systems			
Nature of knowledge	Based on certitude: real objects and their relations are governed by universal laws. Found in science, with Newtonian mechanics and Euclidean geometry as archetypes.	Like true belief, based on scientifically established certitude. Has given rise to such disciplines as risk assessment, toxicology, safety engineering and so on.	Based on certitude, but the certitude is based on such notions as 'naturalness,' 'energy balance' and so on. Often developed into a knowledge system that resembles science (like ecopsychology).
Social progress	Reliance on technological advance to attain social progress	Technological advance also has side effects that need be identified and mitigated. Technological advance only leads to social progress if also social context of technology is adequately shaped.	Social progress requires primarily cultural transition (*Umdenken* = transthinking).
Risks	Social stability under-mined by lack (respect for) expert knowledge and / as a basis for state legitimacy	Unanticipated side effects	Lack of respect for 'natural' and 'historic' standards
Crucial problem dimensions/criteria for assessing action	• Sound scientific basis desired • Need take full advantage of technological potential • Central decision making on the basis of objective knowledge	• Sound scientific basis desired for knowing both effects and side-effects • Take into account impact of social context on utility of technological opportunities • Sufficient plurality of scientific knowledge as a basis for central co-ordination	• Action to be based on systematised knowledge on 'natural systems' • Rely on technology, that fits natural standards • Social action based on collective morale

Such critical assessment on the level of mutual assumptions is, of course, a core element of deliberative political judgement and as such a prerequisite for critically and constructively assessing visions. For example, Shadow Thinkers will suspect that the True Believers' vision of the use of biotechnology to enable more efficient food cultivation would lead to increasing resistence of plants to pesticides and therefore to increasing dependence on ever stronger pesticides. So a critical assessment of the True Believers' vision would include investigating whether or not this claim is valid. Similarly, skepticists would doubt that technology by itself could solve the world food problem, and would question the technological aspects of True Believers' visions from the perspective of the plausibility of the social and political conditions implicitly assumed in these visions.

4.1.2
The Role of the TA Analyst

Another entrance into the methodological implications of the need to regard vision assessment as supportive to political judgement is to focus on the role of the TA analysts. Our point of departure should be the idea that somehow, processes of practical-deliberation-supported-by-TA should offer opportunities for both detached spectatorship and for active involvement. This can be done in a variety of ways, and a discussion of them is way beyond this chapter. Again, the reader is invited to investigate what the following chapters offer in that respect. Let me just offer some clues that may guide that search.

In an insightful article, Bruce Jennings (1987) distinguishes basically three role types. The first one, that of a neutral scientist, he dismisses as deceptive. The other two roles are those of advocate and of counselor. As an *advocate*, the analyst looks at his object from a particular perspective: his own, that of his client/commissioner or whoever. From that perspective he assesses reality and formulates roads for action. Much of the methodology from what we have called elsewhere (Hoppe and Grin 1995; 1999) the 'usable TA' tradition may be used in that case. But we need to add an important qualification. If such a TA is to contribute to political judgement at least two conditions must be fulfilled. First, the TA should be considered as an input in a political process in which also other perspectives are represented. Second, in that process some detached-spectator-type persons should play a role. Not only for the positive reason already mentioned (adding dignity) but also for its negative counterpart: to prevent that the debate "generates more heat than light" as Paul Sabatier (1987) has put it. They may play a role of what Sabatier (1987; with Jenkins-Smith, 1993) calls 'policy brokers,' who help different advocacy coalitions to reach some form of agreement. Returning to methodology, these two 'application conditions' imply that even the analyst-as-advocate should ensure that his arguments are formulated so clearly that they are open to scrutiny by others. The type of expert analysis on robotics, proposed by Michael Decker in his chapter, may serve to illustrate and explore that role and its preconditons.

As a *counselor*, the analyst her or himself plays as much as possible the role of detached spectator, helping more involved actors to reach some form of agreement. As Loeber (1998; Loeber and Grin 1999) has demonstrated, this may be a way to support political judgement as we discussed it above. This role is at the core of two approaches to TA methodology that are being discussed in this volume, especially in the chapters by Decker, Grunwald and Reuzel and van der Wilt. The first is the 'rational technology assessment' approach, proposed by researchers in Germany (Grunwald 1998; this volume). It has been specifically designed as an approach to TA that may contribute to the 'fusion of horizons'. It seems to fit what Dyson (1981) has identified the German tradition of 'rationalist consensus' (*cf.* the comparative study into six national TA's in Hoppe and Grin 1999), and it meets the desire expressed wider (cf. e.g. Gutmann and Thompson 1996) to identify some, albeit non-universal, grounding for deliberation

The analyst as 'detached spectator' is also at the core what we have designated as 'interactive TA,' (Grin and van de Graaf 1996a; Grin et al. 1997) that has emerged especially in countries like Denmark and the Netherlands which are oriented to some form of consensus seeking deliberation (Hoppe and Grin 1999). While it has not been specifically designed for 'fusing horizons,' it may well be used for that purpose. This has been done, for instance, in the Gideon project on sustainable crop protection in the Netherlands, that yielded visions on crop protection in 2030, as well as strategies to realise them (Groenewegen et al. 1996; Grin 1998).

4.2
Two challenges: institutional constraints and long time horizons

We will finish our identification of themes running through this volume by focusing now not on the *implications* of TA as being supportive of political judgement of visions, but rather on its *conditions*. Specifically, the chapters that follow reveal and discuss two major challenges that are involved in such use of TA: the fact that societal institutions have hitherto been shaped by Modern assumptions; and the unusually long time horizon which is, by definition, involved. Let briefly explore these challenges.

First, as Beck (1997, p 58) stipulates, practical deliberation will take place not on some extraterrestrial body, but here, within existing institutions and by people whose basic, often modernist, assumptions will not easily change. This challenge has several aspects, which we will encounter frequently in the pages to come. First, it would seem, 'conversion' on the level of these fundamental presumptions is nearly impossible. Yet, it has appeared possible to identify, from basic literature on learning (e.g. work by Argyris (1976; Argyris et al. 1985) and by Schön (1983)), conditions under which established projects (or, for that matter, visions) may be changed on the level of their basic assumptions (see Sabatier 1987; Sabatier and Jenkins-Smith 1993; Grin and van de Graaf 1996b). Such conditions include severe crisis due to exogenous developments that have an unanticipated impact on established courses of action; and surprises, unanticipated effects of action which may lead to reconsideration of fundamental assumptions especially if these surprises occur frequently and are negatively appreciated.

A different aspect of this challenge may sometimes play a role: a lack of a sufficiently balanced, pluriform knowledge infrastructure. For instance, van der Wilt

(1995) has demonstrated on the basis of an extensive scrutiny of medical professional literature on Parkinson's disease, that 80-90% of all literature reflects just
one perspective on the disease's determinants. Also, concerning military matters,
European countries lack any significant share of expertise outside the military
establishment, and this reduces the plurality of strategic thinking.

As a final, and related, aspect of this challenge, the Cartesian *cosmopolis* as envisaged by especially True Believers (see table 1) has dominated for such a long
time that established scientific knowledge and public, expert-based decision making institutions have in many cases become a 'seamless web.' Examples are the
emergence of 'national innovation systems' (Allen 1994; Nelson 1994) as well as
the development of a discipline like statistics which has co-evolved with governmental institutions for statistical policy research. In the Netherlands, for instance,
the paradigmatic development of the discipline, the emergence of governmental
offices for statistics and the way in which statistical experts judged the validity of
particular policy uses of their work appear to have mutually reinforced each other
(Stamhuis 1989; 1999; de Gans 1999). The same can be said about geodesy,
which in paradigmatic and institutional terms co-evolved with spatial policy planning institutes in countries as different as the United States (Dupree 1963) and the
Netherlands (van der Woude 1987). Thus visions and institutions tend to reinforce
each other, limiting the opportunities for critical assessment and re-construction of
visions.

The second challenge concerns the long time horizon involved. More specifically, it is about the type of control dilemma's referred to in the first section: how
to 'plan' our future, given that we know so little about it; *and* how to prevent that
the visions constructed have the same authoritarian implications as many existing
utopias. One important remark on this issue is that visions that reflect an adequate
mix of (True Believers, Shadow Thinkers, Light Seekers) assumptions by their
nature form a hedge against these risks. This can be easily grasped from table 1,
where it becomes clear that each of these approaches is particularly sensitive to
one particular type of risks and favours its own type of acceptability criteria. A
vision incorporating several of these criteria and risk perceptions combines the
sensitivities of the respective modes of thinking. As Schwarz and Thompson
(1990) have put it: divided we stand.

However, such a hedge is not sufficient. Grunwald argues in his chapter that we
also need to rethink TA from the normative and epistemological questions triggered by our wish to use TA for these purposes. This leads to his proposal of rational TA. Rational TA includes the idea of bringing about a synthesis of different
visions, but it adds important elements to it. 'Rational TA' focuses on a type of
rationality that is not merely substantial, but also procedural. The idea is that vision assessment should be guided by acceptability criteria, not by acceptance.
Also, vision necessarily should be seen as an ongoing process, rather than as a
one-and-for-all result. And it should be about shaping the future rather than about
planning or even forecasting it. The notion of rational TA is explored further in
the chapter by Michael Decker on robotics. The robotics case is an interesting one,
because there hardly exists an articulated vision, at least not in the sense as one
shared by adherents of a variety of disciplines as well societal stakeholders. Thus

Decker uses the case to illustrate some of the challenges involved in actually constructing visions in cases in which visions are still rather underdeveloped.

With the latter phrases we are back where we began this introduction: the idea of shaping the future by continuously adapting the horizons that we follow and that move along with us. It is now time to proceed, and see what can be learnt from the following chapters on ways and means to actually elaborate and apply the methods explored in this introduction. The quest we undertake is as down to earth as it goes *below* the surface. It is an indispensable complement to the epistemological, social-philosophical and macro-sociological work that has been cited and reviewed above, and that remains emptier than is useful unless it is turned into practical prescriptions that help us to enter 21st century in a responsible way.

References

Achterhuis H (1998) De erfenis van de utopie. Ambo, Amsterdam

Allen P (1994) Evolutionary Complex Systems: Models of Technology Change. In: Leydesdorff L, van den Besselaar P (eds) Evolutionary Economic and Chaos Theory: New Direction in Technology Studies. Pinter, London

Andreae JV (1619 [1916]) Christianopolis. Oxford University Press, Oxford

Argyris C (1976) Single and double loop models in research on decisionmaking. Administrative Science Quarterly 21, pp 363-375

Argyris C, Putnam R, McLain Smith D (1985) Action Science. Jossey Brass Publishers, San Francisco London

Beck U (1992) Risk Society: Towards a New Modernity. SAGE, London.

Beck U (1997) The Reinvention of Politics. Rethinking Modernity in the Global Social Order. Polity Press, Cambridge

Beiner R (1983) Political Judgement. Methuen, London

Bernstein RJ (1983) Beyond Objectivism and Relativism. Science, hermeneutics and praxis. University of Pennsylvania Press, Philadelphia

Bernstein RJ (1992) The new constellation. The ethical-political horizons of modernity/postmodernity. The MIT Press, Cambridge, MA

Boeker, E (1975) Natuurwetenschap en techniek: een weg naar Utopia? Van Gorcum, Assen/Amsterdam.

Callenbach E (1975) Ecotopia. Bantam Press, New York

Crombag H, van Dun F (1997) De Utopische Verleiding. Contact, Amsterdam/Antwerpen

de Gans HA (1999) Population forecasting 1895-1945: the transition to modernity. .Kluwer, Dordrecht [etc.]

de Jouvenel B (1963) The pure theory of politics. Cambridge University Press, Cambridge

Dierkes M, Hoffmann U, Marz L (1996) Visions of Technology. Social and Institutional Factors Shaping the Development of New Technologies. Campus Verlag/St.Martin's Press, Frankfurt/New York

Drucker PF (1993) Post-Capitalist Society. HarperBusiness, New York

Dupree AH (1963) Science and the emergence of modern America, 1865-1916. Rand McNally, Chicago

Dyson K (1981) West Germany: the search for a rationalist consensus. In: Richardson J (ed) Policy styles in Western Europe. George Allen & Unwin, London, pp 17-46.

Fischer F (1980) Politics, Values and Public Policy. The Problem of Methodology. Westview Press, Boulder, Col.

Fischer F (1995) Evaluating public policy. Nelson-Hall, Chicago

Fresco LO (1998) Schaduwdenkers en Lichtzoekers. Huizinga-lezing 1998. Bert Bakker, Amsterdam

Gadamer, HG (1975) Truth and Method. Seabury Press, New York.

Grin J (1998) Participation, co-production and power. Rationale and praxis of interactively performed Technology Assessment: the example of the GIDEON project. Paper presented at the International Conference Evaluation: Profession, Business or Politics? organised by the European Evaluation Society, Rome, 29-31 October

Grin J, van de Graaf H (1996a) Technology Assessment as learning. Science, Technology and Human Values 20, no. 1, pp 72-99

Grin J, van de Graaf H (1996b) Implementation as communicative action. An interpretive understanding of interactions between policy actors and target groups. Policy Sciences 29, no. 4, pp 291-319

Grin J, van de Graaf H, Hoppe R (1997) (with a contribution by Peter Groenewegen) Technology assessment through interaction: a guide. SDU, Den Haag (Working document Rathenau Instituut; W57)

Groenewegen P, Reijnen E, van Rijn J, van de Sande T, Grin J, van Laar H, Schreurs C, Reus J, Bouwman G (1996) Op weg naar een duurzame gewasbescherming. Eindrapport. Rathenau Instituut, Den Haag, Studie 35

Grunwald A (1998) Rationale Technikfolgenbeurteilung. Konzeption und methodische Grundlagen. Springer Verlag, Berlin Heidelberg New York

Gutmann A, Thompson D (1996) Democracy and disagreement. Why moral conflict cannot be avoided in politics, and what should be done about it. Belknap Press, Cambridge, MA & London.

Hobsbawm E (1994) Age of Extremes. The Short Twentieth Century, 1914-1991. Abbacus (2nd edition), London

Hoppe R, Grin J (1995) Technology Assessment for Participation, Introduction by the guest editors to a Special Issue on Interactive Technology Assessment of *Industrial and Environmental Crisis Quarterly* 9, no. 1, pp 3-12

Hoppe R, Grin J (1999) Traffic goes through the TA machine. A culturalist comparison between approaches and outputs of six parliamentarian technology assessment agencies' traffic and transport studies. In: Vig N and Paschen H (eds) Parliaments and Technology: the Development of Technology Assessment in Europe. SUNY Press, New York (to appear)

Huizinga J (1935) In de schaduwen van morgen. Tjeenk Willink, Haarlem.

Janich, P (1996) Was ist Wahrheit? Beck Verlag, München.

Jennings B (1987) Interpretation and the practice of policy analysis. In: Fischer F, Forrester J (eds) Confronting values in policy analysis. The politics of criteria. SAGE Publications, Newbury Park etc., pp 128-152

Laursen JC (1992) The politics of skepticism in the ancients, Montaigne, Hume, and Kant. E.J. Brill, Leiden etc.

Lindblom CE (1968) The Policy Making Process. Prentice Hall, Eaglewood Cliffs, NJ

Loeber AM (1998) Dynamiek in beleid door debatten tussen overheid, bedrijfsleven en wetenschappers: het fosfatenbeleid in Nederland. In: Hoppe R, Peterse A (eds) Bouwstenen voor een argumentatieve beleidsanalyse. VUGA, Den Haag, pp 99-114

Loeber AM, Grin J (1998) From green waters to 'green' detergents: processes of learning between policy actors and target groups in eutrophication policy in the Netherlands, 1970-

1987. In: Sabatier PA (ed) An Advocacy Coalition Lens on Environmental Policy. SUNY Pres, New York

Lovelock J (1979) Gaia: a New Look on Life on Earth. Oxford University Press, New York etc.

Lovelock J (1986) Gaia: the world as a living organism. New Scientist, December 18

Mambrey P, Pateau M, Tepper A (1995) Technikentwicklung durch Leitbilder. Neue Steuerungs- und Bewertungsinstrumente. Campus Verlag/St.Martin's Press, Frankfurt/New York

Nelson R (1994) Economic Growth via the co-evolution of technology and institutions. In: Leydesdorff L, van den Besselaar P (eds) Evolutionary Economics and Chaos Theory: new directions in technology studies. Pinter, London, pp 21-32

Nelson R, Winter S (1982) An evolutionary theory of economic change. The Belknap Press, Cambridge, MA & London

Popper K (1963) The Open Society and Its Enemies. Princeton University Press, Princeton

Roszak T (1992) The voice of the earth. An exploration of ecopsychology. Simon & Schuster, New York etc.

Sabatier PA (1987) Knowledge, policy-oriented learning and policy change. An Advocacy Coalition Framework. Knowledge 8, pp 849-892

Sabatier PA, Jenkins-Smith H (1993) Policy change and learning. An Advocacy Coalition Approach. Westview Press, Boulder, Col.

Schön DA (1983) The reflective practitioner. How professionals think in action. Basic Books, New York

Schwarz M, Thompson M (1990) Divided we stand. Redefining technology, politics and social choice. Harvester Wheatsheaf, London

Stamhuis IH (1989) Cijfers en aequaties' en 'kennis der staatskrachten': statistiek in Nederland in de negentiende eeuw. Rodopi, Amsterdam (etc.)

Stamhuis IH (1999) An unbridgeable gap between two approaches of statistics; 1850 the turning point To be published in a volume, edited by Klep PMM and Stamhuis IH on Statistical activity in The Netherlands between 1750 and 1850.

Toffler A (1980) The Third Wave. Pan Books, London

Toffler A, Toffler H (1994) Creating a new civilisation. The politics of the third wave. Turner Publishing, Atlanta, GA

Toulmin S (1991) Cosmopolis. The hidden agenda of modernity. The University of Chicago Press, Chicago

Tuchman BW (1979) A distant mirror. The calamitous 14th century. The Macmillan Press, London

van der Wilt G (1995) Alternative Ways of Framing Parkinsons Disease: Implications for Priorities for Health Care and Biomedical Research. Industrial & Environmental Crisis Quarterly 9, no. 1, pp 13-48

van der Woud A (1987) Het lege land: de ruimtelijke orde van Nederland 1798-1848.Meulenhoff Informatief, Amsterdam

Vergragt PJ, Jansen LA (1993) Sustainable technological development; The making of a Dutch long term oriented technology program. Project Appraisal 8(3) pp 134-140

von Weizsäcker E, Lovins AB, Lovins LH (1997) Factor Four: doubling wealth, halving resource use: the new report to the Club of Rome, Earthscan, London

**II   Case Studies. Technology Assessment and
the Role of Visions**

Technology Assessment as Metaphor Assessment. Visions Guiding the Development of Information and Communications Technologies

Peter Mambrey, August Tepper

1 Introduction

Multiple factors shape the development of new technologies. Very important mental factors are guiding visions (Leitbilder) and metaphors which are used to design technologies or systems. With my colleagues Michael Paetau and August Tepper, I did empirical and theoretical research based on concrete cases (Mambrey et al. 1995). It was our aim to investigate the methodological use of guiding visions and metaphors for technical design and for preventively oriented technology assessment and to suggest adequate tools. Our project combined different research arenas which were previously unconnected. It combined technology assessment (future assessment), genesis of technology (role of Leitbilder in social systems), and linguistics (construction and analysis of metaphors). Two case studies about the role of Leitbilder and metaphors in the development of the typewriter and the personal computer demonstrated the scope of the approach as well as the predictive qualities. A survey on how members of a research and development organization for applied information systems used and constructed metaphors were done.

Despite the large differences between approaches explaining how metaphors work, all implicitly or explicitly share the belief that metaphors open up perspectives and can therefore be used as imaginative mechanisms, similarly to scenarios or other techniques, to prescribe a future. They cannot predict or discover utopia but open perspectives on certain aspects, highlighting them, and obscure others, downplaying them. They are means for communication and enable discourses on technologies which are under construction or even not yet developed (Mambrey et al. 1994). This paper reports on findings how metaphors and guiding visions were used and could be used to assist the design of information and communications technologies (ICT)[1]. The research took place at GMD – German National Research Center for Information Technology, where a large-scale research program on "assisting computers" was begun as early as 1988.

What are the connotations of the vision assisting computers? The goals of the program were to study and develop the principles and methods for assisting sy-

---

[1] The paper is based on research done in the project „Metaphors in Computer Science – Potentials and Risks". The project started in October 1990 and ended September 1993. It was funded by the German Federal Ministry for Education, Research, and Technology.

stems, and to demonstrate assistance properties and capabilities in various proto-type systems. "Research on computers as assistants means looking for new ways of dividing the labor between humans and computers. On the one hand, future systems should take on more tasks than existing systems do, especially those that seem tedious or difficult for humans. On the other hand, they should not automate tasks completely. The basic paradigm is that of assistance. In many fields of application the problems are either too complex or simply too numerous for any attempt to develop a machine with complete problem-solving competence to succeed. What is called for instead is a set of calibrated tools that the user can combine, adapt, and employ as he or she sees fit. Exhaustive treatment and coverage of a problem is in fact not the goal of computers as assistants" (Hoschka 1996, p. 1f). The assistance computer metaphor was a strong guiding vision to communicate specific ideas of the future directions of computing to all developers of the institute. There were many messages built in the notion of assistance in relation to computers. They were all about the relation between man and machine:

- *Hierarchy*: Man is superior than machine. He dominates, he orders and bears the responsibility. This cannot be left to algorithms and chips;
- *Initiative*: Man keeps the initiative to control and advice the machine about its activities;
- *Freedom of action*: Man directs and limits the realm of activity for the machine. The computer is not autonomous, it is dependent on its given goals.
- *Tool*: It is a tool, a machine and not an anthropomorphous entity with its own values.

It is a master-tool relation and does not share the assumption that computers could act and behave autonomously. Within the institute especially in the department of artificial intelligence different views existed about the autonomy of a technical system. Although nobody thought at this time of artificial-life systems some researchers discussed the idea of integrating human capabilities like learning and thinking into autonomous robots. Notions like understanding and intelligence were used to discuss the functionality of future technical artifacts. The assistance metaphor therefore was used to focus the research policy of the institute into the direction of computers as assistants and dependent tools of man.

2   Dilemmas of Technology Assessment

Managers and politicians as well as the public are looking for instruments predicting the risks and benefits of a technology as early as possible. The same problem exists for system designers and users to construct and evaluate future applications. But assessment and design are confronted with dilemmas. These dilemmas restrict the possibility in assessing future systems but they do not make these efforts useless. They are hurdles which ask for new methods (for a detailed review of such dilemmas compare Banse, Friedrich 1996):

- The time and knowledge dilemma: often reliable data are only available when the product has already been placed on the market and nearly nothing can be changed ex post;
- The methodological dilemma: there are no reliable methods to forecast the individual use and societal embedding of technologies, the process of en-culturalisation of technology is unpredictable for very fundamental reasons;
- The steering dilemma: technology cannot be steered by a certain person, by a group of actors, or even by nations presupposing that the steering process would yield exactly and guaranteedly the results intended (Collingridge 1980);
- The judgement dilemma: there is no common judgement on certain tech-nologies, which is unanimously shared in societies. Different perspectives from different points of interest are the reality in the pluriform society. Evaluations are, in the present discussion on TA methodology, mostly re-garded as subjective (for instance, Paschen and Petermann 1991).[2]

Those dilemmas[3] cannot be overrun, but one can search for tools to clarify future development in the sense of evaluating the different orientation frames and make them explicit and open for public discussion. Instead of well-known methods of technology assessment, we applied in our research project a new approach: we directly gained the information from Leitbilder and metaphors used in the first steps of the system design. Our research project was part of a scheme to develop the foundations for a new generation of ICT-based support systems named the "assistance computer". The vision of these new software systems is based on the metaphor of a computer system behaving like good assistants. Our project has refined this metaphor and proved that the analysis of metaphors in computer sci-ence is a valid method for design and future assessment of information and com-munications technologies.

3   Paradigms, Leitbilder, and Metaphors

3.1
What are Metaphors and Guiding Visions?

A basic problem of using visions and metaphors for technology development and technology assessment deals with the meaning of cognitive processes and con-ceptual ideas in nature. It would be relatively easy if our concepts – including visions and metaphors – might directly represent the ontic world existing inde-

---

[2] Compare the case study by Rob Reuzel and Gert Jan van der Wilt in this volume as an example where the evaluations ex ante (done by clinicians mainly) and the evaluations ex post (done by the deaf organisations) did not match.

[3] The dilemma of simultaneously taking into account long-term orientations for technology policy as well as short-ranged flexibility requirements – central to the paper of Armin Grun-wald in this volume – relates to the dilemmas mentioned but adds the aspect of cultural dyna-mics (between continuity and flexibility) to the following consideration.

pendently of our consciousness. Then a vision would not simply be an idea, a cognitive possibility, but contain ontic propositions and would be a directly useful source for any designer (Danesi 1990). However, such an immediate connection is not available, but visions and metaphors are human constructions – as any other concept – that are useful to structure worlds and that are therefore used (Helm 1992; Way 1991).[4] This is sufficient as a basis for understanding here. Frequently made assumptions about ontic qualities of pictures and so on are from our point of view not necessary and also incorrect.

As a tool for analyzing and designing such constructions, three basic concepts are needed: paradigms, guiding visions (Leitbilder), and metaphors. Paradigms and guiding visions resemble each other only at first sight, they should be clearly distinguished. According to Thomas S. Kuhn (1967), paradigms are long-term orientation patterns of science, such as the classical Newtonian mechanics, that were for long periods unquestioned and were reused and proved again and again by "normal" science, as Kuhn calls it. These are the basic selections in a complex world that decide on whether concepts such as guiding visions are considered relevant at all. Here it is only necessary to remember the concept of paradigm as necessary background for a future assessment by metaphors that cannot be questioned directly for lack of an observation position situated at a higher level. Guiding visions and metaphors are like the two sides of a coin. For illustrating different aspects and for the conceptual separation thus required guiding visions denote the function and metaphor denote a frequent and important form of expression. The function of guiding visions is that of a medium that establishes "structural interconnections" between different and largely autonomous social subsystems. By analogy with Talcott Parsons, Niklas Luhmann (1984) called them symbolically generalized communication media, and a very simplified example is public opinion. Public opinion results from an interaction between journalism and politics. Both sides supply news and commentaries, and a great number of only loosely coupled opinions are replaced by a construct to which politics and media strictly adhere. The symbolism that is assigned to the construct is important. Because of their advantages, media such as public opinion or guiding visions are permanently reproduced by the social system as a social instance, namely, by changing and advancing the respective subjects. Dynamics, the change of subjects, is required for the existence of the social instance.

If a Leitbild is expressed already in more or less precise terms, these descriptions can easily be used for discussing technological improvements on a early phase of design to gather orientation. But Leitbilder are relatively seldom expressed in precise terms, especially in the early stages of a new technology. We see two functions of a Leitbild:

---

[4] Remember the various types of thinking about technology and the future like Shadow Thinkers or Light Seekers, mentioned by John Grin in the introduction. Such concepts are, obviously, interpretative constructions and might themselves be designated as metaphors.

- the guiding function: this is a *collective* projection, which can be seen as a synchronization of collective assessments and adjustments;[5]
- the vision function: which is an individual cognitive actuation, an appeal to personality , it is the creative flash of insight, which triggers the actions to develop new products.

Quite often visions cannot be explained by using existing words with a fixed meaning, and people are forced to compose new terms like "paperless office", "knowledge navigator", or "assistance computer". A metaphor is one of the means to produce an utterance of a vision. Therefore the analysis of metaphors used by researchers is not only necessary but much more interesting.

Metaphors are forms of non-literal use of language. Metaphors interactively combine known and new aspects; they explain new aspects in terms of similar ones. The structural and conceptual forces of the metaphors lie in their surreal quality and polysemantic. They indicate a course without tracing it out in detail. It is important, that metaphors are always linguistic entities and not pictures. A picture would be an object about which one can speak with the aid of the metaphors, for example.

Metaphors have a lot of functions, which we analyzed during our research. They are of a high persistence and are often used in a multifunctional way: expressing creative thinking, directing work, allowing communication about an object that does not yet exist in reality and making explanations easier are the most important ones. The functions are:

- creative for the individual;
- communicative for groups;
- marketing for management;
- mnemotechnical and pedagogical within teams;
- establishing boundaries between teams;
- differentiation to other team-works.

To some extent, metaphors thus secure user orientation automatically, because the respective systems are designed within a structure that is more or less familiar to future users. But of course, generally there is no automatism, no omnipotence given by using metaphors for design.

3.2
How do Metaphors Work?

In addition to the question of what metaphors are, it is interesting, how they work. This question is answered differently. Today Max Black's interaction theory is

---

[5] This function of visions – transcending the mere subjective level towards collectively shared goals, aspirations etc. – makes the vision approach attractive to any attempt to arrive at some kind of *trans-subjectivity* in technology assessment (compare the paper of Armin Grunwald in this volume).

widely accepted as a general model of how metaphors work (Indurkhya 1992), but older and alternative theories might explain at least some special cases.

The ancient Greek and Roman philosophers thought that words showed something of the real world (substitution theory) and regarded metaphors as a substitute quite often used only for rhetoric purposes. The task was to discover the "true meaning" behind metaphors.

Today the substitution theory is replaced by more complex theories. Some researchers see metaphors as a comparison transferring meanings from the source to the target domain; for example, an atom is structured like the solar system. Objections based on several mismatches are rejected by restricting the comparison to salient aspects only (ships and planes have captains in common but no falls or sheets). Extracting these aspects allows statements to be made about certain properties of the target domain. Like other authors, Gentner (1982; 1983) restricted the comparison to certain salient aspects. The decisive feature of Gentners approach is the fact that only the relational structure of the source is mapped to the target, not the properties of the relevant objects. In the metaphor "the atom is like the solar system" the relation "revolves around" between sun and planets is true for nucleus and electron too, but the attribute "the sun is yellow" does not carry over. This model is interesting from a methodological point of view because the selection of the salient aspects is supported by a specific category.

Approaches to metaphor research that use pre-cultural explanations pursue a quite different method of problem solving. Examples are Jung, Levi-Strauss, and Lakoff. According to Jung (1976) archetypes are innate psychological structures, ancient patterns of human behavior. Claude Levi-Strauss discovered in metaphors a meta-code that is super-existent for an individual. Lakoff and Johnson pursued an approach that was comparable to some extent (Lakoff 1987; Lakoff and Johnson 1980). They traced back language to the use and combination of basic components (semantic primitives) . These primitives correspond to very original body experiences such as every thing being either inside or outside of the body, and Lakoff and Johnson outlined a metaphorical process combining elementary concepts like components to more complex utterances: "most concepts are partially understood in terms of other concepts" (Lakoff and Johnson 1980, p. 56). Thus, according to their hypothesis, metaphorical utterances can be understood in a relatively simple way by machines and they can be associated with meanings.

Black's interaction theory is the most popular one. Interaction theory is very much connected to the modern constructive philosophy and does not suppose a substitution or a comparison. Black and other researchers see metaphors as the product of an interaction between source and target producing something new unknown before. We do not discover hidden similarities but we construct similarities between different objects by stating and testing such connections. Once you have noticed a new possibility you have mentally created a new object to which people quite often refer with the help of composed words. Metaphors therefore make use of similarities but really name a distinctive new object. Based on this theory, all expressions have literal meanings (either known or new objects), and metaphors can be analyzed in almost the same way as "normal" notions. By interaction, Black (1983) outlined a process that is clearly distinguished from the transference of specific comparison aspects. However, the new interpretation of the metaphorical process still compromises structural analogies as a basis but leads to a new understanding: the projection and construction.

4   Uses of Metaphors

4.1
Metaphors and Guiding Visions as a Clue for Technological Development

We analyzed two historical examples of course not as a proof but as a hermeneutic tool to be inspired about the roles metaphors can play for the technological development. One example was the visions of the personal computer from Vannevar Bush after the second world war, the other was the development of the typewriter two centuries ago.

Vannevar Bush outlined a machine to augment intelligent work and operated by individuals. This vision of a powerful and individually operated computer served as a vision to coordinate the work of different groups at different places and sometimes even different times. The findings correspond to actual research on different topics of explaining technological development (Bijker et al. 1993). The examples show, that guiding visions do play a mayor role in the invention process. In contrast to the idea of the self-emerging evolution of technological development, guiding visions and metaphors enable mutations of technology which are not based on the technical evolution and prospects but on a visionary look forward.

The history of the invention of the typewriter shows that the metaphor "Schreibklavier = writing piano – clavier imprimeur – cembalo scrivano" was well known in several countries and existent in several languages. It played a mayor role in the development process. The metaphor not only made communication about a nonexistent machine possible, but also produced concepts of the later typewriter. This vision contained, for example,

- the keyboard and its layout;
- the power transmission;
- the fixed finger positions;
- the arrangements of the keys and
- the use of all ten fingers simultaneously which enabled greater efficiency (Mambrey 1994).

Using the example of the "Schreibklavier" metaphor and of the two-hundred years history of the "Schreibklavier" as we know it, we wish to cite a few methodological aspects which are important to an anticipatory assessment of future developments by means of metaphors. The analysis point to the persistence of the abstract holistic design concepts and the dynamism of the singular concrete design elements. Thus, if one wishes to utilize metaphors for future technology assessment, it should be done on the basis of a reflection of the abstract holistic concepts and not of individual design elements. Concepts and socially shared ideas appear to have a longer life than design elements. The design elements are frequently sub-jected to change, some of them become obsolete and are replaced by new elements (Typewriter – word processing computer). Ultimately what remains is the pheno-type which represents the design concept, whilst the majority of technical design elements disappear, e.g. mechanical power transmission, physical existence of types.

Since, as figurative speech, the metaphor should not be understood word for word, it follows that we must recognize the core of the metaphor. In the case of the "piano" metaphor the core is the keyboard as a man-machine interface and not the black color for example. This reduces the infinite number of an object's prop-erties. The core of a metaphor itself contains design concepts. Black (1983, p 409) suspects "that many metaphors put us in the position of seeing particular aspects of reality which the creation of the metaphor helps us to constitute" Metaphors are cognitive instruments which are imperative for perceiving connections which, once recognized, then truly exist. It may be the case that no technical design ele-ments are incorporated in metaphors in the sense of the likeness theory (e.g. learning from nature); rather, the metaphor contain concepts which can be per-ceived individually but not verbalized. The metaphor would then be the communi-cations medium for a condensed non-communicable experience. This would ex-plain why one looks in a particular direction – or according to Black – what sector of the reality is actually seen.

For assessing technology we must consider that what is being expressed by a technical metaphor must necessarily be limited to those spheres covered by the metaphor. Using our example "Schreibklavier" this means the man-machine inter-face, and the keyboard in particular. Thus the technical metaphor can express nothing of the need for, or the quantity or quality of this technology, i.e. the way in which society will adapt and incorporate this technology. A technical metaphor is, in contrast to visions, more or less *value-neutral*[6], it is ambigous and polyvalent and does not touch the sphere of societal fears or hopes.

Our example shows that metaphors are media of communication, by means of which the unknown – and thus unnamed new elements can be communicated. They trigger further metaphorical processes through their polysemantics and sur-realism, which in turn support the power of imagination in technological devel-opment and thus have a creative impact.

4.2
Metaphors as Instruments of Daily Work for System Designers

The interviewing of researchers working for different projects within one institute of the GMD – German National Research Center for Information Technology was part of the empirical basis of our studies. The results show that metaphors exist in the daily work of research and design of systems and that they are necessary. We

---

[6] Each terminology contains normative aspects having been invested by making the basic di-stinctions. Strictly value-neutral descriptions, therefore, are not available. However, there is a broad continuum between low and high value correlations included.

found that metaphors were ubiquitous. They are active instruments on at least three hierarchical levels of an organization:

- On the organizational level a global guiding vision of the organization was active which gave the orientation frame to the researchers working on project level.
- On the second level project metaphors enabled communication about the project aims.
- On the third level – the operational level (small groups) or often individual level – metaphors were used as cognitive tools to create new ideas and products.

The different metaphors found on the tree levels coexisted and did not contradict each other.

Often, especially on the operational level of system design, several metaphors were active, creating a closely coupled metaphorical scenario. Very often in order to explain a metaphor to others new metaphors were created. We found in some cases a differentiation process, where existing metaphors were replaced by more "precise" new metaphors, which were commonly understood on interpersonal, project- and organizational levels. So several metaphors coexisted on different levels. This offered the opportunity of focussing on mentifacts or future artifacts as a wide-angle or telephoto lens does. You can also observe them from different perspectives. As Danesi pointed out, images created by metaphors are subject to the same variation parameters that visual percepts are. Some images have a high resolution, while others manifest themselves in fuzzier ways (Danesi 1990). The results show that metaphors are necessary for technical development and are used not in a systematic way as tools but unreflectively as part of the natural language.

4.3
Metaphors as Tools for Requirement Analysis and Design

Our idea was to use metaphors explicitly in a structured way as instruments for orientation and evaluation in technical development. Using metaphors instrumentally is not just playing with words; it is playing with visions, possible targets, future aims, and future solutions. It provides rich information for insight and yields a wider range of perspectives. The instrumental use of metaphors therefore can be to structure and enhance joint brainstorming or self-reflection. The following examples show some structured ways to work explicitly with metaphors, gaining information and orientation in design. They can be used as tools for anticipation and future assessment.

4.3.1
Exploration by Metaphors

Ingendahl (1973) examined the metaphorical use of words for semantic domains and semantic features. According to him, a metaphor is understood by the relation to the content neighbors within the sense domain (paradigmatic relation) and to the neighbors in the context (syntagmatic relation). As a metaphor the word originates from such relations and enters again into such relations. Within the metaphorical process, type and reference points of the relation are changed. These two basic assumptions can be used to reconstruct step by step the meaning of a metaphor. This leads towards a more systematic and explicit use of metaphors. They can be created by applying different approaches. One approach, which we tried out, was to generate and interpret metaphors via linguistic variations. We started with a mother metaphor, divided it into two aspects and differentiated these aspects several times separate from each other. After three steps we combined the separated aspects and identified those which made sense to us.

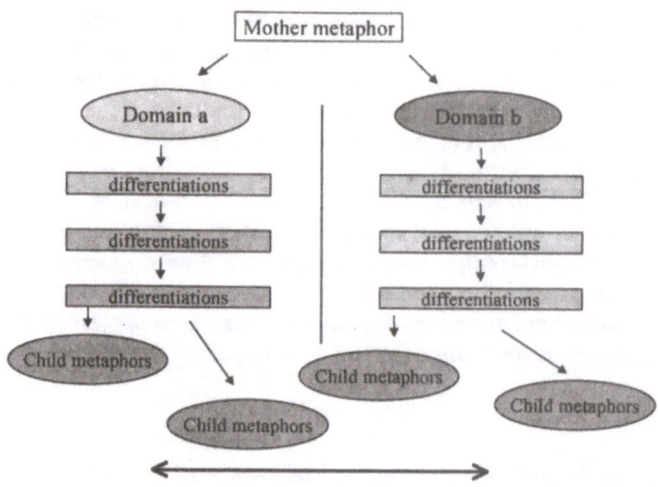

Fig. 1. Generating and interpreting metaphorical scenarios via linguistic variation.

The main problem of this approach is the amount of complexity which it produces. After two or three steps of differentiation a great number of variations could be found which were useful and fruitful for our work. In an open structured brainstorming we jointly generated via linguistic variation names of technical artifacts with its underlying concepts, "task servant" for example.

4.3.2
Classification

Identifying or defining the character of a metaphor made sense as well and let us instrumentally work with metaphors to assist creative thinking. Classifications can be:

- hyperbole
- hypobole
- pars pro toto
- simile
- analogy
- contrast
- litote
- anthropomorphization
- synecdoche
- metonomy
- allegory
- substitution
- change of forms
- incarnation
- etc.

Applying this can help to produce new salient metaphors or insights which enables us to communicate and discuss future technologies.

4.3.3
Semantic primitives

Another interesting approach oriented to the work of Lakoff and Johnson about semantic primitives is the discovery of basic building components in metaphors. Semantic primitives such as container, war and others are not very useful for future assessment because impacts are hard to derive from them. However the elaboration of specific primitives for specific technological areas is most promising. For example, basic ideas for discussing technological impacts can easily be derived from a system component such as "centralization". Each research domain discusses a set of such basic primitives, such as machine or tool. As another example computer science uses primitive categories like evolution, family, and community. Quite often these primitives have fixed meanings (centralization, for example, is viewed as a bad one). The task is to find further and/or better primitives and to analyze the underlying assumptions and concepts for them.

4.3.4
Semantic deduction and choice

Within the framework of our research project "assistance computer" , the first part of this compositum was elaborated with the aid of a thesaurus. In addition to encyclopedias and electronic dictionaries, computer systems able of interpreting metaphors may also be useful (cf. Carbonell 1982). A "true" statement on the word domain will not be attained in any instance, instead the selection and arrangement of the synonyms each constitute special cognitive constructs.

The encyclopedic search for synonyms expands the semantic word domain of a metaphor. As early as the first stage, the verbs "assist" and "aid" provide a number of important alternatives, which, over the course of the following stages, provides clear information on useful or unsuitable properties of an assistance computer. The word assistance can be seen to embrace only a few useful properties (the word assistance derives from the lateen: stare = stand). The most fruitful results are obtained from the elaboration of the word "execute". Forms of support of a more psychological nature, such as "encourage" or "invigorate" are clearly shown to be inappropriate properties of an assistance computer.

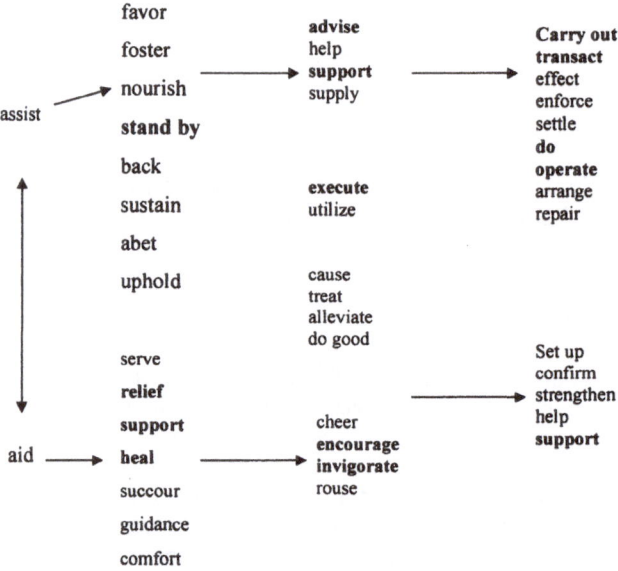

Fig. 2. Elaboration of the assistance computer metaphor into synonyms by a thesaurus and the evaluation of the latter by a group of designers.

One result of this simple description and evaluation process is the fact that a clear demarcation line exists between the properties of human assistance and the desired properties of an assistance computer. The general result is a refined definition of the "assistance" metaphor whose contents have in part been refined and described with greater precision. Such definition refinements are absorbed directly into the systems development process.

4.4
Contrastation and Discourse of Metaphors and Guiding Visions as Future Assessment Tools

In addition to the language oriented assessment approach, the discourses of the prototypes of the assistance computer within the organization has been analyzed. How did we proceed: We changed the focus of analysis. Instead of exploring or constructing meanings of guiding visions and metaphors which are inherited in the mentifact itself we analyzed the ongoing discourses about the Leitbilder in the work settings of our institute. Thus the Leitbild-centered analysis mentioned in the previous sections was added by a discourse-centered analysis. The basic materials of the discourses were the different written papers and project meeting minutes. Their interpretations gave us insight into the concepts and ideas which were behind the metaphors. The previous methods are tools to explore a semantic domain but they cannot inform about the notions and mentifacts being outside of this domain. To overcome this shortage the guiding visions and metaphors were analyzed as objects in discourses, competing with other ideas and guiding visions. They were not seen as standing alone but being embedded in time-dependent discourses. Thus we wanted to identify the implicit assumptions in which the metaphors are embedded in and to identify the changes they underwent during the time of the discourse. Only by analyzing the selected metaphors by observers not being involved in a discourse one can detect important information included or excluded from the design process. The realm of a discourse is much wider than the semantic domain of a metaphor or guiding vision which play the important role to communicate ideas and directions for future design within a group of designers (Foucault 1981, p. 42). Exploring discourses offer more information than exploring single notions. This perspective allows a more extensive impression of Leitbilder and metaphors in related research areas. One can detect what is being deleted and what is brought in as arguments. By this contrastation we could make not only those concepts and ideas explicit which were projected into the existing guiding visions and metaphors but also those concepts and aspects which were excluded. This was a great advantage which occurred via the contrastation. At a very early stage we could make visible which aspects and concepts were part of the engineers' discourses and which were not.

Analyzing the discourse within a specific scientific community offers the chance to contrast the findings with those discourses in other scientific communities. Let us take the example of the vision "design at use". This vision has different discourse arenas. You can find it in the organizational science and computer science as well. Another way of contrastation is comparing the goals. Guiding visions aiming at the same goal can be seen as contrasts, e.g. adaptivity and tailorability as a functionality of a computer system.

Three research arenas of the assistance metaphor were analyzed: Human-Machine Interaction, Computer Supported Collaborative Work, and Artificial Intelligence. Important terms of each of these fields were identified and contrasted with their use in the respective scientific communities. Within these research arenas several scientific discourses were analyzed, for example about:

Organization

The well known understanding of an organization as a machine or mechanism was replaced by a modern view of an organization as a living organism. Organizations were seen as non-trivial machines (Foerster 1993), where different actors and subcultures interact. Self-organization and autopoiesis of the organization was discussed (Luhmann 1984b). So the interest was not to identify stable work-flows of artifacts but to assist cooperation processes in an organization by flexible media and not wired mechanisms. Evolution, tailorability, adaptivity, and flexibility emerged as requirements for a Computer Supported Cooperative System within an organization.

Cooperation and coordination

The designers understood cooperation in an ideal typical sense of "shared goals" and "shared tasks" without reflecting that in other scientific communities cooperation is seen as a special case of the coordination of action by consensus. In general the societal form of production in an industrial society is the industrial plant or organization with a high degree of division of labor which is coordinated not only by consensus, discourse, or trust but by power, norms and rules, obligations, technical processes, and standardized sequences of actions. Instead of focussing the assistance tools to reduce the ill-structured problems, work was done on implementing solutions for well-structured tasks like procurement and the like.

Especially if the Leitbilder were human-centered or technically-centered were from greater interest. The results showed that the assistance metaphor was understood technically. The discourses made clear that within the classification scheme of innovative or non-innovative, which is the dominant polarization of the research and development community of computer science, the understanding was very innovative.

By comparing and contrasting the metaphors and Leitbilder used in a specific context, the interaction between understanding and creation became observable. Because every evaluation needs knowledge about the concrete application context, the task of scientific analysis is as limited as any prophecy. But improved discourses, based on described and structured observations make it possible to discuss consequences and to make evaluations. The three components of an assessment (description/distinction, contrastation, evaluation) can be pursued to any desired degree of complexity.

5   Conclusions

All methods developed for the analysis and the construction of metaphors deliver important information for technology assessment. They explain e.g. the semantic domain of a metaphor. Revealing the semantic structure of a metaphor leads almost instantaneously to an assessment of the components. Take the metaphor of the information superhighway. You can ask, if it is a one-way road, from where to where it leads, who has access, who owns it, who is the stakeholder, what are the requirements for the driver license, how much fine you have to pay and for what, how often traffic jams occur and where ... You can also ask what is excluded and if it is in contradiction to other metaphors e.g. the world as a global village. This short example shows that exploring the semantic domain of a metaphor can construct rich pictures of future developments and make them open for public discussions.

This research could support researchers and developers in better reflecting on the guiding visions and metaphors they use and in revealing excluded aspects. They can be used as tools to reduce complexity and as an orientation frame. They make the discussions about new technical systems more explicit. For example, mediated conferences about guiding visions and metaphors of future technical or organizational developments might be prepared with the aid of these tools. Altogether, future assessment and control based on guiding visions and metaphors is a valuable complement of other methods. However, the relevance of this instrument for technology assessment should not been marred by unreal expectations for its predictive qualities.

No type of prognosis can predict the future reliably. The future is circularly connected with decision-making processes and its results. For the outcome reasonable probabilities cannot be specified at all. The embedding of technological developments into society is a far to complex social process to predict. Therefore technology assessment in general can make only time-dependent proposals whose further fate is open and under discussion. This insecurity is no objection to a guiding-vision-oriented and metaphor-oriented design of technology and technology assessment. But it applies equally to allegedly "more exact" methods. There is no shortcut to utopia. But there are means to construct future technical and social systems which offer a certain amount of control over the process.[7] A technology assessment based on the analysis of guiding visions and metaphors cannot solve the dilemmas mentioned at the beginning of this article, but it can ameliorate the situation with respect to the following aspects:

Knowledge

Guiding visions and metaphors focus on technology-oriented communication processes. The latent meaningful relations inherent in the respective concepts can be revealed, changed, or questioned for possible consequences in the case of their realization. It is a special type of knowledge; it is both comprehensive and polyvalent depending on the specific development phase. Polyvalence is step-by-step reduced by the context. This means that a given selection is consolidated, thus becoming trend-setting.

---

[7] Compare the paper of Michael Decker in this volume dealing with steps of „constructing" a field of discussion on specific robot systems in health care in the early steps of technology development.

Prognosis

Neither traditional technology assessment nor technology assessment based on guiding visions and metaphors can predict consequences in the strict sense (this is concurrently claimed by Armin Grunwald in this volume). Therefore the approach presented here will not change the basic problem. But, perhaps it could contribute to reverse the perspective taken. If the consequences of technology could be predicted exactly, there wouldn't be any free space for shaping technology – prediction implies determinism. Instead, in this paper the perspective of shaping is emphasised, supported by the empirical case studies. In the shaping perspective the task with respect to prediction to dissolve *prognoses* (based on assumptions on social laws) - into *scenarios* (based on decisions at certain points of the development process).

Control

Guiding visions and metaphors are part of the technical development. Any form of more systematic analysis, construction, and evaluation of guiding visions and metaphors is therefore an additional instrument for controlling 'technical development. Changes of guiding visions and metaphors will change technical developments and the adoption process. In spite of this gain of shaping capability the central control dilemma remains. It is impossible to pre-determine the success of a specific technology development.

Evaluation and judgement

Any evaluation first presupposes an object that can be described more or less precisely. In this respect, evaluation of guiding visions and metaphors is dependent on knowledge. This knowledge and the interpretations based on the evaluation of metaphors can lead to a better understanding. This can make hidden aspects overt and accessible to individual judgement or group decisions. But this is only one component of the evaluation process, because evaluation is always a very subjective action and includes power and politics. It is always the question if the realization of the outlined visions are personally desirable or not.[8] A verbal dissection can assist this decision-making process.

Contextualisation

Visions are, on the one hand, dependent on certain context conditions like actor constellations or historical backgrounds. They have to be assessed with respect to

---

[8] Armin Grunwald in this volume attempts to go beyond this subjectivism acknowledging the challenge of the plural society. The most severe problem to solve is to argue upon what basis evaluations and criticisms could be made which could justifiedly claim validity beyond the subjective level.

these specific contextual issues.[9] On the other hand, visions cannot be restricted to specific contexts. The integrative power of visions is that they can integrate different contexts and, by this way, can enforce their impact. Thus visions are situated at a meso-level, between the micro-level of merely singular contexts and the macrolevel of decontextualised societal evolution. This tension between contextualisation and generality has to be observed very carefully in vision assessment.

Participation and cohesion

Metaphors and guiding visions are in general formulated so graphically that the relationship to one's own interests is relatively easy to establish. Their use is important for communicating common goals, finding compromises and social consensus in our societies. They are means to communicate concepts like "holiday society" or "knowledge society". The analysis and discussions of guiding visions and metaphors and the changes effected by that processes are a means of participation of science communities and citizens in public debate on technical and societal goals of a society.

Anticipating the future based on guiding visions and metaphors cannot and should not replace other approaches of technology assessment but complement them. Visions and metaphors naturally have the greatest effectiveness in the definition phase of new technologies (Mambrey 1994). Despite the advantages and disadvantages, no alternative approach is available for this early phase of technology design and technology assessment.

Future assessment by metaphors (Tepper 1993) can be used both as an instrument of self-reflection of researchers and developers and as an instrument of public discussions, for example consensus conferences about distinct metaphors or guiding visions. In later phases of technology design there are alternative approaches. The capability of visions and metaphors to condense diffuse information might advocate their use as tools for future assessment of technological developments.

A caveat

Using metaphors instead of detailed descriptions can be dangerous. Using "wrong" metaphors can lead to "wrong" decisions or actions respective to the use of a technology (Mark and Mambrey 1997). There is no automatism to decide between "right or wrong", there is no proof that defines a salient metaphor in advance, because only practice shows the results. Their context and embedding in societal usage and the individual perspectives define about usefulness or uselessness. In spite of this situation we always have to decide *ex ante* whether we follow certain paths into the future or whether we should chose others. To guide and to improve this reflection ex ante the vision assessment seems to be a suitable means

---

[9] Compare the case study by John Grin in this volume. He uncovers the need for contextual visions assessments in the field of military aircraft use.

– if and only if the unavoidable restrictions of this approach are taken into account by reflection.

References

Banse G, Friedrich K (1996) Sozialorientierte Technikgestaltung – Realiät oder Ilusion? Dilemmata eines Ansatzes. In: Banse G, Friedrich, K (eds) Technik zwischen Erkenntnis und Gestaltung. Edition Sigma, Berlin, pp 141-164

Bijker WE, Hughes TP, Pinch TJ (1993, eds) The Social Construction of Technological Systems. New Directions in the Sociology and History of Technology. The MIT Press, Massachusetts. Cambridge/London. Fourth Printing

Black M (1983) Mehr über die Metapher. In: Haverkamp A (ed) Theorie der Metapher. Darmstadt

Carbonell J (1982) Metaphor: An Inescapable Phenomenon in Natural-Language Comprehension. In: Lehnert WG, Ringle MH (eds) Strategies for Natural Language Processing. Lawrence Erlbaum Associates, Hilsdale N.J.

Collingridge D (1980) The Social Control of Technology. New York

Danesi M (1990) Thinking is seeing. Visual metaphors and the nature of abstract thought. Semiotica 80

Foerster H von (1993) Wissen und Gewissen. Versuch einer Brücke. Suhrkamp Verlag, Frankfurt am Main

Foucault M (1981) Archäologie des Wissens. Suhrkamp Verlag, Frankfurt am Main

Gentner D (1982) Are Scientific Analogies Metaphors? In: Miall DS (ed) Metaphor. Harvester, Brighton, pp 106-132

Gentner D (1983) Structure Mapping: A Theoretical Framework for Analogy. Cognitive Science 7 (1983), pp 155-170

Helm G (1992) Metaphern in der Informatik. Begriffe, Theorien, Prozesse. Reihe Arbeitspapiere der GMD, Nr. 652, St. Augustin

Hoschka P (1996) Computers as Assistants – Introduction and Overview. In: Hoschka P (ed) Computers as Assistants – A New Generation of Support Systems. Lawrence Erlbaum, Mahwah, NJ, pp 1-16

Indurkhya B (1992) Metaphor and Cognition. Kluwer Academic Publishers, Dordrecht

Ingendahl W (1973) Der metaphorische Prozess. Düsseldorf, 2. Auflage

Jung CG (1976) Die Archetypen und das kollektive Unbewußte. Gesammelte Werke Band 9/1. Verlag Walter, Olten und Freiburg

Kuhn TS (1967) Die Struktur wissenschaftlicher Revolutionen. Suhrkamp Verlag, Frankfurt am Main

Lakoff G (1987) Women, Fire, and Dangerous Things. Univ. of Chicago Press, Chicago

Lakoff G, Johnson M (1980) Metaphors We Live By. The Univ. of Chicago Press, Chicago

Levi-Strauss C (1971) Strukturale Anthropologie. Suhrkamp Verlag, Frankfurt am Main

Luhmann N (1984a) Liebe als Passion. Zur Codierung von Intimität. Suhrkamp Verlag, Frankfurt am Main, 4. Auflage

Luhmann N (1984b) Soziale Systeme. Grundriß einer allgemeinen Theorie. Suhrkamp Verlag, Frankfurt am Main

Mambrey P (1994) Die technische Metapher als Kommunikationsmedium und Konstruktionshilfe. In: Jahrbuch Technik und Gesellschaft 7: Konstruktion und Evolution von Technik. Campus Verlag, Frankfurt am Main, New York, pp 127 - 148

Mambrey P (1996) Metaphors as Requirement Analysis Tools. The Market Metaphor in CSCW System Design. In: Shapiro D, Tauber M, Traunmueller R (eds) The Design of Computer Supported Cooperative Work and Groupware Systems. North Holland, Amsterdam, pp. 135-150

Mambrey P, Paetau M, Tepper A (1994) Controlling Visions and Metaphors. In: Duncan K, Krueger K (eds) 13th World Computer Congress 94, Volume 3. Elsevier Science B.V., pp 223-228

Mambrey P, Paetau M, Tepper A (1995) Technikentwicklung durch Leitbilder. Neue Steuerungs- und Bewertungsinstrumente. Campus Verlag, Frankfurt am Main and New York

Mambrey P, Tepper A (1996) Metaphors and Systems Design. In: Hoschka P (ed) Computers as Assistants – A New Generation of Support Systems. Lawrence Erlbaum, Mahwah, NJ 1996, pp. 269-280

Mark G, Mambrey P (1997) Models and Metaphors in Groupware: Towards a Group-Centered Design. In: Howard S, Hammond J, Lindgaard G (eds) Human-Computer Interaction - INTERACT '97. Chapman & Hall, London, pp. 477-484

Paschen H, Petermann Th (1991) Technikfolgenabschätzung - ein strategisches Rahmenkonzept für die Analyse und Bewertung von Technikfolgen. In: Petermann Th (ed) Technikfolgen-Abschätzung als Technikforschung und Politikberatung. Campus, Frankfurt, pp 19-42

Rogers RA (1993) Visions Dancing in Engineer's Heads: The AT&T Quest to Fulfill the Leitbild of a Universal Telephone Service. WZB-Papers FS II 90-102, Berlin

Tepper A (1993) Future Assessment by Metaphors. In: Jorgensen U (ed) Technology & Democracy – The Use and Impact of Technology Assessment in Europe (ECTA III). Proceedings, Vol. II, Copenhagen, pp. 524-537

Way EC (1991) Knowledge Representation and Metaphor. Kluwer Academic Publishers, Dordrecht

Technology Assessment in the Health Care Area: A Matter of Uncovering or of Covering Up?

Rob Reuzel, Gert Jan van der Wilt

1   Introduction

"Franz Vranitsky, a former Austrian Chancellor, is said to have commented that 'anyone with visions needs to see a doctor'", Michael Peckham, co-editor of a book entitled *Clinical futures*, remarks. "It is true he was referring to Europe rather than health. However if the past is anything to go by attempts to predict future changes in medicine are generally wildly inaccurate." Slightly uncovering his own vision, Peckham argues a few sentences later that "the future must see an integration of scientific medicine within a broader framework that tackles social and other determinants of health. This is not an 'either...or' choice but an absolute requirement if there is to be a balanced approach to health development" (Marinker and Peckham 1998, draft chapter 9).

Visions with respect to the world and expectations with respect to technology serve as guidance for developing both the world and its technology. A vision, here, is the world as it is imagined to have developed by use of technology. It is not a prophecy, however, a prediction of what the world will be like, nor is it a utopian fantasy. We would define a vision in accordance with the definition in Chapter 1 as the world described by someone who is asked why particular technologies are desirable. Furthermore, we distinguish visions from perspectives, which are associated with a frame of mind, or a direction of thought, rather than a detailed view of the consequences of such perspectives.

Throughout this chapter, we will be dealing with visions, expectations, and perspectives in relation to concrete technologies. Our main concern is technology assessment (TA), which is defined by Banta and Luce as "a form of policy research that examines short- and long-term social consequences (for example, societal, economic, ethical, legal) of the application of technology" (Banta and Luce 1993, p 61). TA is a tool for valuing particular technologies, and, importantly, the way stakeholders are affected by these technologies. TA seems to reflect that developing the world through technology calls for critical reflection, which implies that underlying visions, expectations, and perspectives are to be explicated and evaluated meticulously.

If this is a task for technology assessment, however, it is not at all clear to what extent this is so. In any case, assessing underlying visions and expectations through technology assessment is easier said than done. For instance, one tends to forget that technology assessment itself is related to visions and expectations, too. "It can be forgotten, in this rational and scientific age, that culture and society underlie all actions in health care," Banta and Luce warn us. This is true of both health technology and its assessment, which "cannot be totally objective or value-free. As an activity carried out by human beings, it too is influenced by social and cultural values" (Banta and Luce 1993, p 132). Moreover, both technology development and assessment may exist in the light of the same visions and expectations. If this happens, the supposed aim of technology assessment as a critical reflection on visions, expectations, and perspectives regarding technology may not always be achieved.

In this chapter, we elaborate on this problem, and explore the potential of TA uncovering visions, expectations, and perspectives that underlie the use of technology in health care. With these objectives in mind, we will try to cover the following ground. First, we address Health Technology Assessment (HTA) in the Netherlands, as it is carried out in the context of one of the major programmes in this area, the Investigative Medicine Programme of the National Health Insurance Council. We will briefly explain this TA programme, and focus on a particular case study that was carried out in its context, viz. the assessment of cochlear implants (CI) for deaf children. Second, and with respect to this case study, we will address the following questions:

I.   *Did* the TA help to uncover and critically examine visions, expectations, and perspectives underlying the technology?

II.   *Should* TA, generally speaking, be used as such, and, if so, why?

III. *Can* TA be used as such, and, if so, how, which methods and procedures should be employed that may be conducive to this end?

2   The Investigative Medicine Programme of the National Health Insurance Council and its advisory task to the Minister of Health

The Investigative Medicine Programme of the National Health Insurance Council is a response to the problem of allocating (scarce) resources in health care, which, roughly, since the sixties has been most pressing in health care policy. Therefore, assessing health technology in the framework of the Investigative Medicine Programme essentially involves calculating the effectiveness or benefit to cost ratio of one technology compared to another. The Investigative Medicine Programme entails an annual budget of approximately 18 million ECU for this kind of research in the field of health technology. As such, it is the largest HTA-programme in the Netherlands. Furthermore, it is comparable to the NHS R&D programme in the UK, and the programme initiated by the AHCPR in the US. Moreover, it reflects what mainstream HTA entails.

Proposals for research mainly originate from university hospitals and schools of medicine. Of these, a few are selected on the basis of policy relevance and methodological rigor, and awarded the financial means to conduct the research. Especially novel technologies with potentially major budget implications are selected. Existing technologies of questionable value are less often scrutinised.

On the basis of project results, the council advises the Minister of Health with respect to eligibility, inclusion of the service into the benefit package, etc. Major projects recently conducted concern the national programme for breast cancer screening, and solid organ transplantation programmes. In the following, we elaborate on the case of cochlear implantation in prelingually deaf children.

3   The Case of Pediatric Cochlear Implantation

3.1
Cochlear Implantation

Cochlear implantation (CI) is a technology that has been developed to provide a sense of hearing to profoundly deaf persons. That is, for persons who cannot benefit from conventional hearing aids based on amplification of sound. Basically, the technology takes over the function of the ear by receiving sounds, translating these sounds into electrical signals, and transmitting these signals to the auditory nerve. The procedure consists of inserting an electrode into the cochlea, and connecting this electrode to an outer transducer. After implantation, several years of intensive training follow, during which a recipient should learn to interpret sound and acquire oral communication skills, that is, produce and perceive oral language.

According to advocates of CI, the technology is to help integrating deaf people into the 'hearing society'. Deaf would be less dependent on sign interpreters and video-based information, be better able to attend mainstream schools, get jobs, and so forth.

3.2
Assessment and approval

The technology obtained FDA-approval for use in postlingually deaf adults in 1984, and for use in pre- and postlingually deaf children in 1990. Purpose of the assessments that were carried out was to obtain information about safety (peroperative complications, rejection of the foreign body etc.), efficacy (notably oral communication skills), and costs.

In the Netherlands an assessment of CI was performed as part of the Investigative Medicine Programme in the term 1993 till 1996. This was an observational study. The researchers came to the conclusion that the CI-team that carried out the implantations in the Netherlands produced results that were comparable to those reported by CI-teams from other countries.

On the basis of the results of the investigative medicine project, the Dutch Health Insurance Council concluded that the technology was no longer experimental, that it had therapeutic value, and that it featured an acceptable cost-benefit ratio. In short, the council argued that CI should be included into the benefit package. However, the Minister of Health disagreed. Various organisations had approached the Minister, and expressed their concerns regarding the technology and a bias in the assessment. The Minister consulted the Health Council, and decided that CI for children should, as yet, not be included into the benefit package. Furthermore, she argued that further studies were needed.

3.3
Responses from Deaf Organisations

Responses from deaf organisations to the assessments were, and still are, not enthusiastic. Why? Their main argument is that deafness is not a handicap at all, but should be regarded as a cultural feature of a linguistic minority using sign language. Deafness in a sociological sense 'refers to socio-cultural characteristics of those hearing-impaired persons who consider themselves to belong to a special (Deaf) community' (Tellings 1995, p 21). The perspective on deafness as a handicap to be eradicated is thought of as a threat to the deaf culture. In addition, it testifies of a low opinion of the deaf by the hearing majority. Especially deaf children would be in danger of experiencing social and emotional pressures, due to discrimination and high expectations, the effects of which could be serious and lasting. Instead, many deaf prefer to be skilled users of Sign language, rather than poor users of oral language. Therefore, they strive for formal recognition of Sign language (in which, till now, they have succeeded only in a limited number of countries), and other measures to support life in the deaf community. Furthermore, there has been some outrage about the resource use associated with CI, and concern about the public funding of other services for deaf people. In short, deaf organisations challenge the perspective on deafness and CI that had surfaced in both development and assessment thus far, and hold the expectations of advocates of CI to be unrealistic.

The British Deaf Association (BDA) in its formal report on CI maintains that it 'could not, as yet, recommend CI for prelingually deaf children' as such 'required an assessment where alternatives to CI are given an equal opportunity of demonstrating success' (BDA 1994). According to the BDA, alternative visions, expectations, and perspectives should be translated into research questions, too, and be inquired into accordingly. Since this has not been done in previous assessments, these assessments suffer from (normative) bias.

3.4
Current Situation

The debate on CI has evolved over the last decades, and to us it seems that the different perspectives have somewhat converged. Currently, in the Netherlands, there is almost no one who is absolutely against CI. Even most deaf agree that at least to acquire some 'survival speech' is likely to improve their life. From this it can be concluded that there is an important difference between CI as an artefact and CI as a socio-technological development. CI as an artefact might be acceptable to all persons involved, if societal consequences of CI as a socio-technological development are acceptable, too. According to deaf organisations, this requires that the vision of a normalised, i.e. hearing, society as a leading vision be replaced by one in which deaf culture is sustained and respected as a full alternative. Indeed, the debate on CI currently centres on the *conditions* of CI being acceptable to all persons involved.

4   Visions, Expectations, and Perspectives – Part I: What is behind CI?

Now, which visions, expectations, and perspectives can be identified in this case? Advocates of CI have been accused of promoting a 'deafless society' in a fashion that reminded of the persecution of Jews, homosexuals, the deaf themselves, and others during World War II, but such a vision is not widely shared (Tellings 1995, p 141). The 'deafless society' has never been promoted as the aim of CI. It cannot be denied that it is a consequence of routine use of CI, though. However, from the perspective of physicians, this possibly is a consequence of *helping* deaf individuals, not of a vision of something like a 'deafless society' being realised.

We wish to emphasise that the vision of a 'deafless society' is a philosopher's reconstruction, rather than a vision that really has influenced the history of CI. By this, we do not refer to the reconstruction of a professional philosopher, but of a researcher who, with hindsight, tries to make sense of a particular history by re-telling the story in a simplified and personalised manner. As we will argue later, we *do* want to know the reconstructions of stakeholders. However, philosopher's reconstructions that stem from a researcher's imagination are the opposite of what we wish to achieve when uncovering the visions, expectations, and perspectives that underlie a particular technology. Measures should be taken to prevent technology assessment from being biased in this respect. On the other hand, we cannot help interpreting things from a particular framework. Therefore, if TA is used to uncover underlying assumptions, this uncovering procedure should include a check with persons involved.

Otorhinolaryngologists did not adhere to a vision of the 'deafless society.' In our experience, they did not have a detailed vision of the world with CI at all. They had a perspective on deafness as a handicap, though, and therefore expected to help deaf individuals through implantation.

Implicitly, they assumed that helping means normalising. As such, the medical culture in which CI was developed is typically modern in the sense Michel Foucault has described 'modern': reflecting a kind of rationality, emergent from the seventeenth century on, that allowed for the creation of abnormal and undesirable by establishing the standard for what is normal and desirable (Foucault 1975). Therefore, the modern notion of rationality is, in a sense, discriminating; it enables the distinction between normal and abnormal.

This modern rationality is associated with the rapid change of health care during the last four or five centuries. A major shift, in our opinion, entails health care having become a professional affair. Accumulation of scientific knowledge and methods more and more require specialized education and skills. To a large extent, knowledge about health is now attributed to physicians, and limited to the physician's domain (Reuzel 1999). Until, roughly, the eighteenth century, however, 'health' was a social denominator: it was related to a sense of wellbeing in a social context. Currently, health is established by professionals. This is not a matter of physicians placing health care out of context, but of health being defined as something physicians can deal with. As a consequence, health and wellbeing have become distinct qualities, wellbeing being more firmly rooted in personal experience.

Notwithstanding the changed definition of health, however, we continue to see health as our highest good as if it still is a synonym for wellbeing in the old sense. Therefore, we continue to leave health care to professionals, and easily accept their care as appropriate.

Foucault has described the transition to 'modern' health care in *Birth of the clinic*. In this book, he shows that the vision of a modern, so-called rational, health care has not been superimposed on existing practices in a sudden act of enlightenment. Rather, this vision has emerged slowly as a result of many minor changes carried out in response to problems met then, and easily overlooked now.[1] When placed alongside *A history of folly*, it becomes clear that the kind of rationality that, we think, characterises modernity is related to a progressive institutionalisation and professionalization of health care.[2]

The works of Barbara Duden prove illustrative here. In her book on the eighteenth century German doctor Storch, she shows convincingly how things change slowly (Duden 1991). From a scientific point of view, Storch had good reason to doubt the account his patients themselves provided of their illnesses. However, his professional knowledge did not, and could not, overrule these accounts. The patients consulted a doctor, but the duty to care for themselves remained theirs, them being part of a social environment. Only later, probably not until the twentieth century, did physicians acquire the power to take over health care.

Following the line just drawn, we would argue that the perspective on deafness and CI that has been dominant in recent years is a professional perspective. This is a perspective on health as something abnormal having been dealt with, that is, a perspective on normalised health. Deaf subjects becoming normal – that is the kind of expectation that has guided the development of CI (Montgomery 1991). However, in the case of CI the expectation that health in the modern sense promotes wellbeing is being challenged. Consequently, the perspective on deafness as a handicap is not shared by all.

The interplay between technology and its underlying visions, expectations, and perspectives is still more complex. For instance, it would be interesting to know

---

[1] When we recognise visions, we should be careful not merely to narrate and summarise the past in the light of our visions. Again, we should beware of philosopher's reconstructions.

[2] Note that the modern concept of 'rationality' described here has been much critisized recently, and contradicts the post-modern understanding of rationality that can be found, albeit tacitly, in our discussion of interactive technology assessment, later in this chapter, and which is elaborated on more extensively by Armin Grunwald in this volume. Currently, however, medicine still hinges upon a modern kind of rationality.

whether the development of CI has given rise to societal pressures that lay as a heavy burden on the shoulders of parents facing the choice for or against CI for their children. For CI may well have changed the perspective on deafness. Inevitable as it was in earlier times to remain deaf when deafened, it has now become a choice, for which responsibility must be assumed. This renders parents susceptible to public pressure. So, apart from the definition of deafness as a feature of those belonging to a deaf community, there is another sense in which deafness is defined sociologically, namely deafness as something that meets more or less acceptance in society.

Our point is that, in this sense, deafness is defined through CI. Visions, expectations, and perspectives do not only *influence* technology development, use, and assessment. Visions, expectations, and perspectives are *shaped* by technology, too. For this reason, assessing assumptions underlying technology is a difficult task. In history of CI, however, no attempt was made at all.

5   Did TA Help to Uncover Visions, Expectations, and Perspectives?

Clearly, in the case of CI in deaf children, the objective of the assessment undertaken in the framework of the Investigative Medicine Programme was *not* to uncover propositions behind the technology; it did not take the critical perspective. Rather, by adopting the professional perspective, major assumptions remained unchallenged and were reinforced through the assessment. So, our answer to the first question, 'Did TA help to uncover and critically examine visions, expectations, and perspectives underlying the technology?', is clearly negative. The TA was carried out in the light of the same perspective as had guided the development of CI in the first place.

Now, a legitimate question would be whether this case was rather exceptional, or whether our conclusions have broader significance. To be sure, the case of CI in children is exceptional, in the sense that assessments have rarely given rise to such public debate. This is partly related to the emancipation of the deaf, which involves a long history of struggle, and to the fact that deaf people and parents of deaf children are relatively well organised. However, other examples exist, where parties involved did not share the dominant perspective that was adopted in the assessment. Various prenatal screening techniques, for instance, have aroused debates similar to the debate over CI. An example from outside the health care area is provided by studies of poverty in Dutch society: in recent years, there has been some debate on the definition of 'poverty' in these studies (Peper 1998).

Wider generalisation from our findings is also possible in view of the central role of clinicians in technology assessments. Usually clinicians take the initiative for this type of studies: they select the subject of evaluation, that is, the evaluand. They also play a key role in conducting the study, because these studies involve patients and patient care. To the clinicians, the investigative medicine fund offers an opportunity for funding the use of novel technologies. Therefore, it is small wonder that their perspective is dominant in most assessments.

Common belief in facts adds to this. Many of us believe that physicians doing scientific research provide objective data, or facts. It is not that physicians should be distrusted; the point is that the belief in facts is a modern, rational faith. At least, it reflects the professional vision on health care. Also, data collected through TA are commonly regarded as facts, which means that they are supposed to be indisputable from a scientific perspective. However, the case of CI illustrates that facts are not value-free. They result from underlying visions, expectations, and perspectives being reflected in choices with respect to criteria of merit, standards of performance, etc. These are normative choices, because they mirror what is considered relevant for subsequent decision-making, and good for patients. Especially when decision-making implies that research findings are straightforwardly translated into policy, it can easily be imagined that facts are not value-free. Still, facts appeal to decision-makers, and to many others, too, as they are thought to facilitate, rather than complicate, decision-making. For decision-makers, it would be most easy if facts *implied* decisions.

Of course, this is not realistic. Facts do not imply decisions, as if they are value-free and clear beyond need of interpretation. However, as an ideal, the search for value-free facts can be recognised in health TA. Apart from the wish for an unmediated link between facts and decision-making, the fear of moral perplexity frustrating progress might be an important reason for this. Moreover, alternative perspectives on TA, such as reflected in participatory approaches, require different methodologies, and perhaps even research infrastructures different from the way health technology assessment is currently organised. We would carefully suggest, however, that, on balance, alternative approaches might be appropriate in some cases. In the next section, we explain why.

6   Should TA, Generally Speaking, Help to Uncover, and, if so, Why?

TA does not always help to uncover visions, expectations, and perspectives underlying the development and use of technology. But does it matter? Our answer to this question is two-fold. It matters, if there is a pretension, a claim of objectivity. Mainstream HTA holds that assessment is an essentially value-neutral research activity, and that value judgements enter only at the stage of subsequent decision-making. Only at that stage, it has to be decided whether, given what a technology can do, and given what it costs, it should be part of the basic benefit package. If this value-neutrality is claimed, against tenability, then it certainly matters! For it implies imposing a particular value framework – one which is not being made explicit, cannot be scrutinised for its validity, and tends to reinforce existing inequities in access to decision-making processes. This part of the argument could be qualified as internal critique: actual practice of mainstream HTA does not live up to the self-opinionated standards of objectivity.

Our second argument could be qualified as an external critique, and is contingent upon a normative and preferred way of decision making that is closely tied up with a particular epistemological and ethical view. For the case study also helps to show that HTA is not a matter of collecting *the* facts about a health technology, as is often claimed. Rather, it is a matter of collecting facts about a health technology that are considered meaningful, given a particular value framework, plausible, given a particular body of knowledge, and amenable to empirical research, given particular methodological standards. This refers to what Schön coined 'overarching theories' (Schön 1983), reflecting particular value frameworks, bodies of knowledge and methodological standards. As we know, these are not always fully shared, and the question is, how can evaluative researchers deal with this epistemic and ethical pluriformity? As was phrased by Smits and Leyten:

> 'The positions and interests of groups which have few opportunities of their own to produce scientific foundations, can be sufficiently interesting and important to explore and develop more fully' (Smits and Leyten 1991, p 25).

Uncovering is still a task for TA

Still, we do favour researchers in the field of TA critically examining visions, expectations, and perspectives regarding technology, rather than decision-makers, ethicists, or others. The argument, here, is not that the perspective of TA researchers should be dominant. Rather, we argue that, first, TA should be comprehensive in the sense that different aspects of a technology and various perspectives on this technology are evaluated in a single assessment. Second, we think that specialised researchers are needed, who as co-ordinators, moderators, and methodological experts are able to cope with the complexity of health care technology in its social context. It might be difficult to repair the validity of assessment results afterwards, as would be the case when different professionals outside the HTA area dealt with explicating visions, expectations, and perspectives separately. This is especially true for assessments suffering from normative bias as a result of adopting a rather lopsided perspective.

Nevertheless, decision-makers, ethicists, or others may assist during the assessment. For they certainly have something to offer, and allowing them to contribute to the assessment may improve thoroughness of ethical deliberation as well as the TA's relevance and appropriateness for decision-making. Professionals and lay people from outside the field of TA are likely to bring their own visions, and have their own expectations and perspectives. As these visions, expectations, and perspectives contrast with each other, the task of identifying visions becomes easier.

Indeed, it might be best to design assessment procedures in which various persons involved participate. In the next section, we elaborate on this idea.

7   Can TA Help to Uncover, and, if so, How?

We think that, yes, TA can help to uncover and critically examine visions, expectations, and perspectives underlying the development and use of technology. But how? Which methods and procedures should be employed that may be conducive to this end?

7.1
Imagination and Consultation

First, we would subscribe to what sounds like, but certainly is not, a platitude: evaluators, use your imagination! Evaluators should always do so, in order to at least avoid the pitfalls provided by routine, and answer most adequately on basic questions like what is it I evaluate, why, for whom, etc., before choosing for a particular evaluation design. However, using your imagination may not be enough. Even experienced researchers may be overcharged. For instance, what is the credibility of issues that merely stem from an evaluator's imagination? What is their relevance? It might be necessary to attract researchers from different fields, or educate practising researchers to improve their capability to examine underlying visions and expectations.

An obvious way to proceed is wide consultation. What are the views of deaf individuals and of parents of deaf children on the technology? In what respect do they differ from those, expressed by developers of the technology and otorhinolaryngologists? What can we learn from this? Which premises seem to be shared, and which not, and what implications could this have for the assessment of the technology?

Wide consultation would result in a shopping list: physicians wish to know whether a technology works, patients wish to know how this technology might affect their lives, policy makers wish to know how much it costs and how its use can be constrained, etc. This is one argument in favour of interactive approaches in evaluation research: different stakeholders may have different information needs, based on their perspectives on the technology. These stakeholders may considered to be different clients of the evaluator. We would acknowledge that they might have *legitimate* information needs. Accordingly, the evaluator should aspire to meet these needs, increasing both relevance and value of the evaluation to a broader audience (Stake 1986).

In the case of CI, we have tried to identify and articulate some of the contentious issues. Let us give you some examples.

Either or

An assumption that since long has been made is that communication in oral language and communication in sign language are mutually inconsistent trajectories: you either raise a deaf child in an oral tradition, or you raise it in a sign tradition, but you cannot have it both ways; the development of one of these systems will go at the expense of the other. We are unaware of any evidence to support this proposition. Indeed, the opposite may be true, in the sense that both systems are mutually facilitative. Examples exist of deaf children who have been implanted and who improved in their ability to communicate in Sign language. To be sure, this is not an intended effect of CI; moreover, the evidence so far is anecdotal. The reason for this is, of course, that such an outcome does not fit the framework of expectations that underlay the development of the technology and its subsequent assessment. Therefore, there has been no research into the subject.

Deafness and Identity

To subject a child to cochlear implantation may also be held as evidence that the child itself is not accepted; there would be not only a rejection of deafness, but a rejection of the child as a deaf child as well. Therefore, there is concern among parents of deaf children that CI may produce feelings of inadequacy and low self-esteem, especially if the children become adolescents: 'What is wrong with deafness, and what is wrong with me, that justifies the substantial and unremitting effort to conform to the auditory majority?'

What is being tested? Validity of Causal Inferences

How can we reliably assess the effects of CI, and what assumptions are behind these assessments themselves? To understand this issue, imagine that we are asked to learn to communicate in Sign language: in an experimental room with no possibility of distraction, at a distance of two meters, a skilled user of Sign language sits in front of us. This person communicates to us in Sign language, and our task consists of replicating the signs. Upon replication, an audience of skilled Sign language users should indicate the meaning of the Sign, as they understood it. The proportion of correctly understood Signs is the outcome of this test. Now, what is the meaning of this proportion? Does it reflect our skill as a Sign language user? Does it tell something about the intervention (courses to learn Sign language)? Not necessarily. Obviously, if we had poor vision, our score would be low, but the test outcome would tell more about our vision than about the quality of the courses in Sign language that we had. The same holds if we had some motor dysfunction, which impaired our ability to produce signs accurately. But even if our vision and motor function are all right, a good test outcome still need not imply that we are skilled users of Sign language. Indeed, we may be completely unaware of the meaning of the Signs we make! We could transmit them correctly, without having any understanding of their meaning.

The testing of the effects of CI in deaf children is prone to similar potential biases. The same room for testing is used, giving no distractions, visually or auditory (no background noise). The task consists now of reading words or sentences, and pronouncing them. There is a panel of listeners, who try to understand what you say. The proportion of correctly understood items is, again, the outcome. Does this reflect the subject's ability to communicate in oral language? Again, not necessarily so. The task requires reading, and many deaf people have great difficulty in reading. Also, the person may produce the sounds correctly, without properly understanding their meaning, or without being able to use them in day-to-day situations (such as school, family, work, etc.).

Note that in this type of studies, there is always a causal inference, a hypothesis to be tested. Conclusions are never merely descriptive ('The children in this study performed so and so before the implantation, and so and so 12 months after the implantation'), they take on the form of a causal inference ('In the children in this study, *CI caused* such and such, etc.').

Inconclusive evidence complemented with expected results

Even if the empirical evidence regarding the efficacy of CI is incomplete and inconclusive, surely the data suggest that it works? These are the outcomes that are expected. This expectation is based on the fact that the CI takes over the functioning of the ear in normally hearing individuals. But does it? What do we know of the neurophysiology of the auditory system, and its rather overwhelming complexity?

7.2
Interaction

Apart from using ones imagination and wide consultation, a third option exists, namely to bring about interaction among stakeholders. The objective of interaction would be not only to elicit different frameworks and perspectives on a technology, but to work towards some kind of agreement with respect to the technology, too. Agreement would entail such questions as 'What are relevant issues to be addressed in the course of an HTA?' or 'What are the conditions of including a technology into the reimbursement package being acceptable to all stakeholders?'

Wide consultation procedures may reveal that in certain instances, there may be a multiplicity of propositions behind a technological development, which may be mutually inconsistent, and none of these need to be exclusively right, at the expense of all other propositions. The question, here, is when we should merely make multiplicity in value frameworks (appreciative systems) and cognitive frameworks (overarching theories) explicit, or whether we should also try to establish a certain degree of agreement among stakeholders. More specifically, should we try to establish agreement among stakeholders concerning the relevance, plausibility and feasibility of issues to be addressed in a technology assessment? Is this the evaluator's responsibility? And do we get better technology assessments in this way? What is agreement? How do we know it is there? Is there anything beyond agreement? Can you expect stakeholders to commit themselves to the results of the process of achieving agreement? It seems to us that these might be important issues for future agendas.

Besides, what happens to the issues that are addressed in TA? Is it always possible to assess their relevance and plausibility, or even to assess their truth? What happens to the issues not addressed? It seems that they remain conjectural, and, therefore, of little relevance to policy-making. When is this undesirable?

If we make multiplicity in value frameworks and cognitive frameworks explicit, and discover various perspectives on a technology being incompatible, then the question is raised whether there is some form of relativism possible, without getting caught up in nihilism. That is, is sensible debate or negotiation about the issues at stake still possible?

Finally, what about closure of the whole process? Is it still possible, and feasible within limits of time and budget, to conduct assessments that, as a result of different stakeholders participating, tend to become rather comprehensive and difficult to manage?

7.3
Rough Outlines of an Interactive Approach

We cannot yet claim to be able to answer all the questions mentioned above. Moreover, to do so would lead beyond the scope of this paper. However, from the things said so far it can be inferred that any evaluation methodology that is promising in this respect features the following points.

1. First, it is to acknowledge that different perspectives on a technology are associated with different information needs. If these information needs are stakeholders' needs, that is, needs of persons affected by the technology, its assessment, and subsequent decision-making, then they ought to be considered *legitimate*.
2. Second, justification is based on acceptance by all persons involved, that is, a kind of agreement. Here, we assume that the value of a technology is the value stakeholders attribute to this technology, not a universal value. This is a pragmatic assertion, agreement being regarded as a local attainment, and the problem of conflicting perspectives being approached as a lack of agreement. Accordingly, technology assessment is a local exercise: it is about valuing technology, but *relevant* values do not stem from an aggregate level.
3. Third, technology assessment should have a constructivist nature. That is, in some cases, it should help designing desirable technology, more than judging technology with hindsight. Positivist inquiries, such as the mainstream health technology assessments currently made, may provide useful and valid information, but cannot determine the (pluralistic) value of a technology. Positivist inquiries can be useful, but should be regarded as tools for decision making, rather than evaluations. For they do not allow, or with great difficulty only, for taking a step back and reconsidering crucial choices being made in the course of the assessment. It is better to consider these choices and adapt the evaluation design beforehand, allowing different stakeholders to contribute. The most promising approach to acceptable technology is to *constructively* incorporate

different stakeholder perspectives in designing technology before we are over-taken by developments.

4. Fourth, it should include some element of fruitful debate. Fruitful debate is always possible, and useful as to make existing visions, expectations, and perspectives with respect to a technology explicit, even if disagreement regarding the technology is not resolved, and agreement is not achieved. This is what Weiss has called 'a revisionist assumption' (Weiss 1983, p 86).[3]

5. Fifth, it should be supported with rules of closure that render the interactive process feasible. Here, we cannot deal with this issue extensively. However, closure in interactive technology assessment involves such aspects as, for example: modesty with respect to goal setting, limits to the number of stakeholders involved (Grin *et al.*, 1997), methods for judging the relevance and plausibility of arguments forwarded, making clear from the outset what kind of decisions are going to be made on the basis of the evaluation, etc.

8   Visions, Expectations, and Perspectives – Part II: What is behind Interactive TA?

As we have said, technology assessment is itself a product of its era. This remains true, even if we better succeed in developing procedures that allow for visions, expectations, and perspectives to be explicated and scrutinised in the future. In this section, we will provide a few additional remarks about the place of health technology assessment in its cultural-philosophical context. Which visions, expectations, and perspectives underlie health technology assessment?

Mainstream health technology assessment, as said, serves as an aid in confining expenditures in health care. Here, visions of a society in which citizens have fair and equal access to health care can be recognised in the background. Accordingly, technology assessment was expected to be an aid in setting fair limits to the costs of health care. The perspective on health technology assessment as a 'science of valuing' (Shadish *et al.* 1995, p 74) beyond cost-effectiveness was there, but up to now has not been translated into adequate procedures.

Since we have related the medical, professional perspective to a modern kind of rationality, one might be tempted to relate alternatives to health technology assessment to times when the modern rationality had not yet developed. This kind of nostalgia can be found in Toulmin's plead for 're-contextualization,' which entails a return to the scepticism of Montaigne. That would be unrealistic, however. Interactive approaches like we have suggested, cannot be regarded as a 're-contextualization.' Certainly, an TA that meets the given requirements can neither be regarded as a Cartesian system of thought as it does not strive for universal knowledge, even not on the basis of local reality and the method of inductivism.

---

[3] We would add that debate *in itself* is worthwhile, if only to prevent us from exclusively adhering to a scientific rationality that is poor for its emotional underdevelopment.

Rather, such a TA is an exponent of post-modern thought. The balance between universalism and scepticism is found in coherentism. Furthermore, truth is a local and temporary construction, which is relevant only in the light of its potential to support action. This adds an important sense of pragmatism to the character of post-modern TA. Moreover, it replaces the vision of TA as a means of assessing technology in view of *subsequent* decision-making. In post-modern TA evaluation and decision-making are merged. Judging is no longer emphasised; instead, nego-tiating and achieving agreement are.

Therefore, post-modern TA is based on another vision, namely the vision of a deeply rooted democracy. In this democracy, people are empowered to influence technology development in an open political system that features shared decision-making and instant access to relevant information. At the same time, people would have to show prudence as Aristotle defined it: the skill to find the middle between seeming opposites. That is, they would be expected to deal with various trade-offs, such as the trade-off between short-term acceptance and long-term acceptability (Grunwald, this volume), individual preference and societal good, ideal state of affairs and political reality, etc.

In recent years, this vision has been translated in evaluation methodologies. One example is Guba and Lincoln's *Fourth Generation Evaluation*. Also, recent attempts to revitalise 'the lost tradition of narrative' (Greenhalgh 1999) should be seen as rays of the brighter sun this vision promises. Both methodologies are based upon the ideas that (a) the truth of the matter is in the stories of persons involved, and (b) in a group effort it is possible to reconstruct shared stories that make sense. Kundera, a novellist, probably smiles at his *L'art du roman* thus be-ing promoted! However, there are still many problems to be tackled. It leads be-yond the scope of this text to work them out in detail.

9   Conclusion

So where does all this leave us? We have tried to show that in this particular case, TA did not assume the critical role of uncovering tacit assumptions. On the con-trary, it largely acted to reinforce these assumptions, by giving them a certain degree of objectivity. We tried to argue that this is a reason for concern, both on the basis of internal and external critique. We finally proceeded by trying to show that an interactive approach might help the evaluator to uncover crucial assump-tions. However, many questions remain, some practical, some more fundamental. It seems to us that we have not yet succeeded in developing a formal method to identify and make explicit (let alone scrutinise) these assumptions.

Interactive TA represents a hopeful development in this respect. Further, works like Schön's *The reflective practitioner* and Rossi and Chen's theory-driven evaluations are certainly helpful. Their message to professional evaluators is that we should start by trying to reconstruct why the evaluand is likely to work. What causal assumptions are being made? What is the validity of these assumptions?

When assessing visions, perhaps an important question should be added: why does one adhere to a particular causal model, apart from its validity?

Indeed, we are trying to incorporate this technique in our daily work, but our experience so far is not an unqualified success. The bottom line is that you are drawing attention to gaps in knowledge and understanding, to tacit and unchallenged assumptions, etc., and the client (in our case the medical expert) is rarely grateful for this.

What is more, in current practice of health technology assessment, notably within the framework of the Investigative Medicine Programme, physicians take the initiative to set up an assessment. Which implies that these assessments are set up to serve their interests and information needs in the first place, and that the hypotheses tested are based on the causal model provided by the physician.

Of course, this causal model can be challenged from points of views other than the physicians', or anyone's. Cases like the case of CI provide strong arguments for doing so. However, this involves a radical transition from the positivist science of testing hypotheses and pre-ordinate assumptions to the constructivist science of shaping the evaluand and its assessment in a group effort. As a consequence, evaluation and decision making and evaluation are merged, justice in decision making becomes a kind of procedural justice, and existing power distributions are altered (cf. Habermas 1983). New procedures must be established to adapt evaluation research to these changes.

To be sure, these are major changes, not easily dealt with. We should take the possibilities seriously, however. Even with our visions being assessed, some dreams remain.

References

Aristotle, Ηθικα Νικομαχεια

Banta HD, Luce BR (1993) Health technology and its assessment; an international perspective. Oxford University Press, Oxford

Chen H, Rossi PH (1987) The theory-driven approach to validity. Evaluation and program planning 10, pp 95-103

Duden B (1991) Geschichte unter der Haut – ein Eisenacher Arzt und seine Patientinnen um 1730. Klett-Cotta, Stuttgart

Foucault M (1976) L'histoire de la folie à l'âge classique. Gallimard, Paris

Foucault M (1963) Naissance de la clinique – une archéologie du regard médical. Presses Universitaires de France, Paris

Greenhalgh T, Hurwitz B (1999) Why study narrative? BMJ 318, pp 48-50

Grin J, van de Graaf H, Hoppe R (1997) Technology assessment through interaction: a guide. SDU, Den Haag (working document Rathenau Institute; W57).

Guba EG, Lincoln YS (1989) Fourth Generation Evaluation.SAGE, Newbury Park, CA

Habermas J (1983) Moralbewußtsein und kommunikatives Handeln. Suhrkamp Verlag, Frankfurt am Main

Kundera M (1988) L'art du roman. Gallimard, Paris

Marinker M, Peckham M (1998) Clinical futures. BMJ Books, London (draft)

Montgomery G (1991) Bionic miracle or megabuck acupuncture? The need for a broader context in the evaluation of cochlear implants. Perspectives in Deafness. A deaf American monograph, February/March 1991, pp 97-106

Peper B (1998) Sociale problemen en de moderne samenleving: een cultuursociologische beschouwing [Social problems and modern society — a treatise in the sociology of culture in Dutch]. Het Spinhuis, Amsterdam

Reuzel RPB (1999) Van klaarblijkelijk tot MRI: MRI in de geschiedenis van de medische blik [From obvious to MRI: MRI in the history of the medical view in Dutch]. Tijdschrift voor Geneeskunde en Ethiek 9(1), pp 7-12

Schön D (1983) The reflective practitioner; how professionals think in action. Basic Books, New York

Scriven M (1984) The logic of evaluation. Edgepress, Inverness, MA

Shadish WR, Cook TD, Leviton LC (1995) Foundations of program evaluation: theories of practice. SAGE, Newbury Park

Smits R, Leyten J (1991) Technology assessment: waakhond of speurhond [Technology assessment: watchdog or tracker – towards a comprehensive technology policy in Dutch].Kerckebosch bv., Zeist

Stake RE (1986) Quieting reform. University of Illinois Press, Urbana

Tellings A (1991) The two hundred years war in deaf education (thesis). Faculty of Social Sciences, Nijmegen

Toulmin S (1990) Cosmopolis; the hidden agenda of modernity. The Free Press, New York

Weiss CH (1983) Toward the future of stakeholder approaches in evaluation. In: Bryk AS (ed) Stakeholder-based evaluation, New directions for program evaluation. Jossey-Bass, San Francisco

Bloodless War or Bloody Non-Sense?
Technology Assessment's Role in Uncovering
Old Propositions behind New Airpower Concepts

John Grin

1   Introduction

Whoever has, since the end of the Cold War, had a glance in strategic studies and military professional literature, undoubtedly has been struck by the enormous amount of attention paid to the so-called Revolution in Political and Military Affairs (RPMA). This revolution is claimed to be a very profound one, more than, for instance, the rapid advance of airpower since the first manned flight in 1903 and its impact on warfare.

One of the most influential contributions to the current debate has come from Alvin and Heidi Toffler. In their book *War and Anti-War*, they claim that we are witnessing the emergence of a new era of warfare. Third Wave War, as they call it, is rising from the ashes of second wave war that was "launched by the Industrial Revolution" (1994, p 44). Its adagium was to maximise destruction. By now, second wave warfare has proceeded "beyond the absolute," with mass destruction as "the deadly *Doppelgänger* of mass production" (1993, pp 48, 50). This realisation, they hold, has triggered a revolution toward third wave war, a new generation of war that reflects the ongoing changes in western economy: towards information as the key element driving economies, and corresponding managerial and organisational changes.

According to Alvin and Heidi Toffler, Third Wave War basically is a war characterized by de-massification and informatisation. Part of the rationale is that the American public "[expects] quick victory and abhors unnecessary casualties" and "reserves the right to reconsider their support should any of these conditions not be met." (US Army Field Manual (FM) 100-5 (1993), quoted by the Tofflers (1994, p 68) with clear approval). To understand their argument, it is furthermore necessary to point out that they foresee, in the 21st century, "collisions of war forms" (1994, pp 104-110). While the United States and other western countries will increasingly fight according to third wave concepts, their adversaries in many cases will be following second wave or even first wave prescriptions.

To be sure, while it can be seen both from the content and from the references of airpower literature that the Toffler concept has had a tremendous impact, it has also met fierce criticism.[1] Critics basically hold that the historical analysis on which the Tofflers base their argument is superficial at best and that (therefore)

---

[1]   For a good review of these criticisms, see Bunker (1996).

their proposals for the future are flawed. Yet, at first sight at least, it is difficult to suppress the suspicion that even strong and thoughtful critics like Bunker (1996; also 1997) appear to be still guided by the same Big Basic Bias that has been guiding the Toffler proposal and its less critical receptions. More precisely, virtually all (semi-)official concepts for 21ˢᵗ century warfare seem to be based on the same principles that have governed defence planning throughout the 20ᵗʰ century. If this is correct, there is *a priori* reason to assess to what extent current 'revolutionary' concepts are prone to the same problems that have been articulated earlier regarding previous concepts.

In this paper, focusing on airpower, I will attempt to uncover these assumptions and, subsequently, to use that analysis to critically assess airpower concepts for the 21ˢᵗ century, in such a way that also the contours of conceivable alternative concepts can be identified. The underlying objective is to learn about the ways and means through which TA can contribute to critical political judgement on attainable futures – our *futuribles*, as futurologist Bernard de Jouvenel (1963) has called them. My attempts to do so depart from the premise that it is our – *a posteriori!* – construction of the past, that shapes our views of the future; and, therefore, that a critical review of concepts of the past has relevance for assessing concepts for the future.

In the remainder of this chapter, I will first explore the emergence and substance of the dominant horizon of expectation guiding airpower planning from 1903 through its canonisation in two recent, influential airpower publications. Subsequently, I will argue in section 3 that the 'revolutionary' concepts now being proposed for the 21ˢᵗ century, basically reflect the same horizon. This raises the suspicion that they may suffer from similar shortcomings as those brought forward between 1975 and 1989 against then existing airpower technology and strategic and operational concepts. So as to be able to verify or falsify this presumption, I will then, in section 4, outline what approaches were taken in technology assessments in which these technologies and concepts were critically investigated. This approaches will then be used in section 5 to assess concepts for 21ˢᵗ century airpower. The paper will end with a discussion of how and where alternative concepts may be identified and elaborated, and how TA may contribute to that exercise.

2   The 20th Century Vision Guiding Airpower

2.1
The Emergence and Entrenchment of an Established Vision

Renowned military historian Martin van Creveld (1989, pp 183-187; see also Gunston 1978) has the history of military aviation start in 1783, when the first hot air and hydrogen balloons were built and used. By the end of the same year, a tract had been published in Amsterdam with a wide range of speculations concerning the military use of these 'flying globes,' in particular to capture Gibraltar from the English. Shortly after that, during the Napoleontic wars, pamphlets were spread containing detailed descriptions of the use of balloons in the military, including their use to transport units of the French Army across the English channel. In the nineteenth century, balloons were actually used during wars, and this showed not only their possibilities but also their limitations, in particular those implied by operational circumstances (such as unfortunate and unexpected weather circumstances). As a consequence, they were only used for siege warfare or other stationary circumstances.

Yet, by the end of the nineteenth century balloons had become an established part of the army's force structure, and strategic thinkers kept dreaming up new expectations, in which they were not only used for reconnaissance, but also for dropping explosives with high speed at significant depth. This must, of course, be seen against the background of the strong and hardly contested faith in technological progress to solve practical problems that, at the time, was expanding very rapidly and very widely (Ter Borg 1985; Van Lente 1988; Toulmin 1991). Parallel to these developing strategic expectations was technology development, guided by the desire to make balloons dirigible so as to overcome their limitations. Given these strategic considerations and these directions of technology development, it should not surprise us that once manned, powered and thus dirigible aircraft had been invented, they virtually immediately were used for exactly these two missions. In 1911, only eight years after the first manned, powered flight by the Wright brothers, they were used by the Italians for reconnaissance and bombing against the Turks during the invasion of occupied areas in Lybia.

The Italians' experiences on their turn inspired both further technology development and strategic and tactical thinking, initially especially by the Germans. In the area of strategic and tactical innovation, it was soon discovered – although unexpectedly to some (Morrow 1996) – that the bombers and reconnaissance & target acquisition aircraft that originally represented the only military missions might, under circumstances, meet adversary fire, both from the air and from the ground. Thus in the early years of World War I, not only the idea of air combat was being developed, but also the requirements of manoeuvrability started to enter tactics. More precisely, the tactical requirement was to be able to outmanoeuvre the adversary: 'do the same but do it better'. So on the one hand, technology development was guided by the idea of increasing speed, agility, altitude, range and payload; and on the other hand, conceptual development aimed at some sort of 'absolute' tactical superiority. Thus technological and conceptual expectations were mutually reinforcing. Concepts called for and reckoned on better technology; technology development aimed at and was believed to realise these wishes, *and* to overcome existing technological and operational limitations.

This particular combination of tactical and operational expectations reflected the visions of strategic thinkers since 1783. At the core of this vision is the ideal of significantly increased speed, agility, altitude, range and payload of air vehicles. This vision inspired conceptual planners to plan types of warfare based on tactical and strategic superiority. It inspired technological planners to work on a variety of technologies to increase airpower's 'inherent attributes' of speed, acceleration, range and payload. Underlying this vision is the belief that technological advance will ultimately make all this possible, in spite of - or, may be better: compensating for - operational limitations. We will now see, from some important examples of technological and conceptual airpower developments, that this vision and the underlying faith in technological progress have been guiding airpower planning ever since.

Engine Improvements

Our first example, improving engine thrust and thrust-to-weight ratio as well as improving fuel efficiency, very soon became strong 'guiding principles' (Smit 1989) in engine development. For instance, while the Wright brothers still had a thrust-to-weight ratio of 1:15, engineers are now working towards a 15:1 ratio. These trends were originally not motivated by tactical and strategic considerations. Already in 1912, before the first experiences with air combat had taken place, the German *Kaiser* launched an award for the best engine meeting ambitious thrust, thrust-to-weight and fuel efficiency requirements, simply because the properties of then existing engines seriously limited the very possibility of flight, at least of flight with some endurance (Gunston 1978, p 77). Yet, improved thrust and thrust-to-weight imply improved speed and acceleration, and thus improved manoeuvrability. And, supported by the belief that technological advance would bring more, this inspired additional, even larger expectations concerning manoeuvrability. Conceptual and technological development thus soon started to reinforce each other around the vision of increasing speed, range and payload. The result was a continuous improvement of thrust, thrust-to-weight ratios and fuel efficiency over time along with conceptual developments that posed ever increasing demands on agility, altitude, penetration depth, speed, payload and so on.

These lasting, intertwining technological and doctrinal trends are not self-evident. While increasing thrust and thrust-to-weight ratio were initially undertaken to make endured flight possible at all, it is obvious that later advances can no longer be explained from the original motive. Similarly, that manoeuvrability entered tactics was initially an unavoidable response to the emergence of air warfare; but this still cannot explain the idea of *outmanoeuvring* the adversary, a notion that logically implies the need to be similar but better, and therefore absolutely superior.

What these long standing trends therefore illustrate are deeper convictions. More specifically, they reflect the just mentioned vision and the underlying belief in technological advance. The technological developments that have increased thrust, thrust-to-weight ratio and fuel efficiency over the past few decades reflect the quest of maximising *all* inherent airpower attributes: speed, agility, altitude, range and payload.

This can be easily seen, first, from the fact that improvements in thrust, thrust-to-weight ratio and specific fuel consumption directly led to improvements in speed, agility and altitude. Second, and still more indirectly, especially improvements in thrust-to-weight ratio and fuel efficiency have made it possible to maximise capabilities in terms of *all* attributes, without having to accept that one improves one attribute at the cost of reducing another. Most obviously, this is because these features imply increased range and/or payload: less fuel is needed for the same range, so that more payload can be taken or a better range is possible. But at least as interesting is that it made design compromises easier (see the example in Box 1) (Grin 1992, pp 231-233). We may conclude, therefore, that it has been the quest for ever improved inherent attributes that explains these trends.

> **Box 1:** *For instance, interceptors demand a low specific fuel consumption at high speeds to realise a combination of high penetration speeds and long range. This requires a bypass ratio of at least 1 in the engine, since only then a low cruise fuel consumption can be combined with good acceleration and take-off capabilities. For air combat fighters, on the other hand, a bypass ratio of less than 0,5 is preferred. They are used at near-maximum thrust most of the time, and to avoid a high fuel consumption at that level, a much lower bypass ratio is needed. Improved thrust-to-weight ratios as well as improved fuel efficiency have made a compromise possible: a bypass ratio somewhere between 0,5 and 1, without unacceptable loss of range and payload capabilities.*

Increasing Complexity of Overall Design

A second, perhaps more imaginative illustration of a development trend guided by airpower's classical vision and the underlying belief is the increasing complexity of military aircraft design. The relation between the just discussed trends in engine design and the trend of ever increasing complexity has been summarised concisely and sharply by the British air force historian Guy Hartcup (1993). First, as he puts it, "the development of the jet engine was leading to the design of faster and faster aircraft, each bringing with it new problems of handling and control, problems that the electronic scientist was rapidly overcoming with a new complex of electronic aids" (1993, p 289). In addition, "(...) so great was the speed at which strike aircraft now operated in their nap-of-the-earth mode that such aids as terrain-following radar and flight control and weapons delivery computers became essential ingredients of the avionics of the strike aircraft" (1993, p 290).

These quotes make clear that the increase in overall complexity is not just a matter of increasing numbers of parts per aircraft, although this type of increase in complexity has also been very impressive. To mention but one example: while the GE J-79 engine employed in the well-known F-4 Phantom contained only 999 parts, the Pratt&Whitney F-100 engine used in the F-15 and F-16 contained some 4541 parts (Binkin 1986, pp 44-45). But Hartcup's claims are referring to something further going: rapidly increasing *functional* complexity. That his suggestion that this phenomenon can be explained by the trends toward increased speed, agility, altitude, range and payload capabilities is surely plausible, can be seen from two examples.

The first concerns 'active control technology'. Traditionally, dynamic stability required that the tail of aircraft had a negative lift; clearly, from the viewpoint of speed and payload, this was disadvantageous. Active control technology makes inherently unstable designs possible. Rapid automatic feed-back mechanisms translate off-equilibrium excursions of an unstable aircraft into correcting movements of control surfaces. This relieves designers from rigid requirements on dynamic stability. In turn, this enables either increase of total lift of the aircraft, or reduction of lift surfaces and associated drag and weight, leading to reduced engine power requirements. This implies better range and/or payload: the saved engine weight and improved lift may be used to take more fuel or more ammunition. Also, it relieves the tension between controllability and dynamic stability, thus improving agility and altitude (Walker 1987, pp 30-32; Mason 1987, pp 7-9; Gunston 1984, pp 63-66).

Another example is the concept of Mission Adaptive Wings, introduced as a response to the fact that the speed enabled by improvements in thrust and thrust-to-weight ratio can only be exploited by stronger wings. In principle, this requires heavier wings, thus reducing thrust-to-weight ratio. Mission Adaptive Wings are shaped, depending upon the amount of lift and drag required for a specific stage of a sortie, through tens of moving surfaces at the forward and trailing edges of the wing. All these parts are controlled in real-time by sophisticated, costly avionics, thus contributing to complexity (Walker 1987, pp 23-24).

Thus the trend toward ever increasing complexity has been caused to a significant extent by the desire to improve *all* inherent attributes, benefiting from engine improvements.

To be sure, there are other factors contributing to increasing complexity. Hartcup's analysis shows but one of the two categories of explanatory factors that have been revealed by Mary Kaldor in her classical work on the "baroque arsenal". The other category is competition: between different industries, between different services and governmental agencies and between different countries. Kaldor (1982, p 15) herself mentions the well-known example of the Multi-Role Combat Aircraft (MRCA), more popularly known as the Tornado: "Britain wants MRCA for long-range strike and strategic air defence (against bombers). Germany wants MRCA for close air support. Italy wants MRCA for air superiority (against fighters)." Thus emerged what the German defence analyst Ulrich Albrecht (1974; cited in translation in Kaldor 1982, p 15) has called an "egg laying, wool-producing, milk-giving sow". It is easy to add many other examples. For instance, in the British Tactical Strike and Reconnaissance project in the late 1950s, competition between the RAF, the Royal Navy, the Treasury as well as the desire of many firms to be involved in the project ultimately led to a design that, compared with alternative proposals formulated during the process, could hardly have been more complex (Law and Callon 1992). Similar processes appeared at work in the course of the development of successive generations of US strategic bombers (Enserink 1993), as well as during the development of the European Fighter Aircraft (Enserink et al. 1992).

Yet, this does not answer the question *why increasing complexity is a guiding principle* governing competition in aircraft projects. For instance, as we will see below, increasing complexity generally leads to increasing costs. If the cooperative nature of aircraft projects is one of the causes of increasing complexity, how can we understand that international co-operation like in the MRCA project is undertaken as a way to deal with these cost increases? This seems highly irrational! It is *not*, however, if we consider the type of choices on overall design made in specific projects. In projects like those just referred to, the problem is generally defined as: how to design a new generation aircraft - that is an aircraft *technologically* superior to that of the previous generation - and deal with the problem that it will be significantly more expensive? And the answer formulated is generally: benefit from technological advances either to get at least the same amount of combat power per coin unit, or to design an aircraft that can play several roles (e.g. both ground attack and protect itself, thus reducing the amount of interceptors needed to protect the fleet). In other words: underlying the trend toward increasing complexity eventually is the strong belief that the strategic problem of meeting strategic objectives with finite budgets can be met through exploiting technological advance.

Thus we conclude that technology development has been significantly shaped by the guiding principles of maximising speed, agility, altitude, range and payload. Second, we have seen that we can only understand the prolonged time over which as well as the specific ways in which this quest has been pursued from the belief in technological advance. It is this belief that encourages designers to be not content with just some improvement in one attribute, e.g. speed, but to work towards maximising all of them; it is this belief that makes it rational for the actors involved to maintain the trend toward increasing complexity; and it is this belief that is responsible for the lack of consideration of operational limitations. The central role of this belief becomes even more plausible when we turn to the area of conceptual planning. Here one might expect that smart tactics and operations are put central, rather than technological advance. Yet, over the past decades airpower concepts have emphasised the latter as their focus.

The FOFA Operational Concept

Let us consider an example from a period of conceptual innovation, the early 1980s, when NATO countries were concerned about the perceived numerical superiority of the Warsaw Treaty Organisation (WTO) land forces, and the rapidly decreasing public acceptance of using tactical nuclear weapons to deter such an attack. Given resources constraints, it was not possible to deal with the perceived threat through increasing the number of weapon systems and manpower. NATO countries concluded that they should exploit their technological superiority. In 1985, NATO's Defence Planning Committee published the so-called Conceptual Military Framework (CMF), that was supposed to guide conceptual planning and technology development by and among member countries. In the wording of this 'vision,' high technology was seen as a 'force multiplier' to off-set quantitative numerical superiority. Conceptually, the core ideas were to "seize the initiative" and "to keep manageable the force ratios at the forward edge of the battle area." Specifically, this meant beating the adversary through relatively mobile forms of manoeuvre warfare, as well as attacking Follow-On Forces before they would reach the battle area. The latter was not only to keep force ratios manageable at the place of direct engagement. In addition, it was seen as contributing to denying the initiative to the adversary through breaking the momentum of the ground attack and disrupting his operational plans.

Of course, Follow-On Forces Attack was not a new mission. Interdiction was already a priority mission in General Dwight Eisenhower's 1943 US Army's Field Manual 100-200, *Command and Employment of Air Power* (Bingham 1996, p 30). New was that interdiction was now seen and elaborated as from a technology-as-a-force-multiplier philosophy. New was also the idea to engage, in particular, *moving* troops (OTA 1988). In the years following the Rogers proposal, a consensus emerged between European NATO countries and Rogers' staff that the emphasis should be on attacking moving troops at depths up to 150 km from the forward line of own troops (FLOT). New was also the type of technology that would be necessary. The most important challenges concerned systems to acquire and engage such moving targets.

Thus what we see is an operational concept that not just emphasises technological advance to better meet strategic and operational requirements. Much further going, the idea is to attain strategic objectives through an operational concept that implies considerably stronger operational requirements and assumes that they can be met technologically. Even more directly than the technological trends discussed above, this reveals the underlying belief that technological advance *per se* is able to solve strategic and operational problems. The idea is to make strong technological progress, and to adopt operational concepts that fully exploit what then becomes possible. The conceptual and technological vision laid down in the CMF reflects this central and fundamental belief. Operational capabilities are reduced to technological capabilities; operational limitations are hardly considered.[2]

2.2
The Canon gets Codified: 'Ten Propositions Regarding Airpower'

The CMF's vision is typical for the type of visions that have dominated western air forces. It reflects the fundamental airpower canon that has emerged since early this century. At the core of this canon is the idea that technology does not just enable planners to better fulfil given operational and strategic requirements, but that technological progress *per se* is the central asset: not merely technological but also conceptual planning should be focused on fully exploiting the expected technological potential.

Recently, an interesting and probably influential attempt has been done to codify this canon: the *Ten propositions regarding Airpower*, formulated by Colonel Phillip Meilinger (1995). According to the author, the ten propositions are an attempt to catch the thrust of air force theorists' approaches to airpower in a set of propositions with sufficient heuristic power to guide the future air force. These emphasise the good old airpower attributes: speed, elevation, lethality and flexibility. Meilinger stresses that these attributes had been thought up even before the military aircraft had been invented. We have seen that this view is quite plausible.

---

[2] These assumptions have been revealed as typical for US defence planning generally by Spinney (1980, p 120).

On the other hand, Meilinger stresses, there have always been some inherent weaknesses of the air force. Airplanes cannot live in their medium, and thus need to land for refuelling, maintenance, repair and rearmament. They were hindered by bad weather and the night. And, finally, airplanes are not able to seize and hold ground.

In a brief epilogue, Colonel Meilinger notes that, for a long time, airpower strategic concepts expected more of technology than technology could meet. However, he claims, "airpower has now passed through its childhood and adolescence, and the wars of the past decade - especially in the Persian Gulf - have shown that it has now reached maturity." (1995, p 70) That is, the ten propositions reflect expectations regarding airpower that have by now become reliable heuristics, generating more specific guiding principles for future airpower concepts and technology.

The underlying assumptions become particularly clear from comparing the substance of these expectations with his list of airpower attributes and weaknesses. Even a quick look is revealing: his ten propositions are rooted in airpower's attributes rather than in its inherent weaknesses. The attributes - speed, elevation, lethality and flexibility - concern the basic *technological* capabilities of aircraft. The argument is that in our days technology has proceeded so much that classical expectation can, finally, be met. The limitations, on the other hand, include operational circumstances: the need for landing, bad weather and night. The limitations also include politico-strategic considerations, deriving from the fact that airpower by itself cannot seize and hold territory.

Box 2: Ten Propositions Regarding Airpower

1. *Whoever controls the air generally controls the surface.*
2. *Airpower is an inherently strategic force.*
3. *Airpower is primarily an offensive weapon.*
4. *In essence, airpower is targeting; targeting is intelligence; intelligence is analyzing the effects of air operations.*
5. *Airpower produces physical and psychical shock by dominating time.*
6. *Airpower can simultaneously conduct parallel operations at all levels of war.*
7. *Precision air weapons have redefined the meaning of mass.*
8. *Airpower's unique characteristics require centralized control by airmen.*
9. *Technology and airpower are integrally and synergistically related.*
10. *Airpower includes not only military assets, but aerospace industry and commercial aviation.*

How widespread these assumptions are becomes even more clear when we consider the so-called 'twelve principles emerging from the ten propositions.' The principles have been suggested by Colonel Richard Szafranski (1995) as a critical comment on Meilinger's propositions. To be sure, in spite of their friendly tone his comments are critical and cutting deep. Szafranski emphasises the fact that airpower cannot conquer the ground on which we live, and thus concludes that only joint, inter-service operations can be of strategic relevance. In addition, he emphasises operational circumstances such as clever tactics and operational behaviour of the adversary, and he emphasises that to be successful, "technology must be applied within superior concepts of operations and codified in superior doctrine" (1995, p 79). In these important respects, his more nuanced views do undoubtedly more justice to historical experience than those of Meilinger.[3]

Box 3: Twelve Principles

1. *A proposition can be an assertion, not a proof or a truth.*
2. *Control the height or pay the prize.*
3. *Airpower can be a peculiarly strategic force.*
4. *Strike the enemy to create opportunities.*
5. *Airpower is about applying force to nodes, processes, webs, intersections, and unions.*
6. *Enemies are bound to be resilient.*
7. *Combined arms aim at convergent effects.*
8. *Mass is concentrated force.*
9. *The object of force application determines the form of force control.*
10. *The informed application of superior technology can vitiate the enemy.*
11. *Technology is unconfinable.*
12. *Effective integration can produce superior force.*

Yet, at some more basic level, the two Colonels' assumptions are the same. Szafranski too seems to accept the well-established notion that strategy and operational concepts can be based on merely technologically inspired expectations. While this is clear from his discussion of his twelve principles generally, it is particularly clear from the ninth principle, which suggests that operational concepts should be shaped such that they enable to exploit most of superior technology. That is, they assume that the best way to proceed in airpower planning is to maximally benefit from technological advancements to exploit airpower's 'inherent attributes.' The ten propositions and the twelve principles thus reflect the same assumptions as those underlying the type of vision we have met in section 2:

– they depart from technological opportunities;
– they lack serious consideration of operational limitations; and
– they reflect a 'symmetry principle:' do basically the same as potential adversaries, but do it better (and better and ...) - either by better technology, as Meilinger would emphasise (and is typical for many aircraft programmes fo-

---

[3] That this is true is plausible, for instance, from the different accounts the authors give of the experiences during the Second Gulf War. Note that this does not necessarily imply that Meilinger's claim is equally wrong that at least some of his propositions give an accurate summary of the *thinking* of a wide range of airpower theorists of this century.

cusing on agility, altitude with the promise that this equals manoeuvrability and therefore survivability), or by an operational concept that is better able to exploit technological superiority, as Szafranski would preferably have it (and of which the FOFA concept was a fine example).

The main distinction with earlier visions is the greater emphasis on more recent attributes, especially high precision.

3   The Assumptions Underlying Concepts for 21st Century Warfare

Is it correct, as I suggested in the introduction, that the just discussed canon underlying earlier airpower visions has also been guiding concepts marking the so-called revolution in political and military affairs? In order to answer this question, we need to discuss concepts for 21st century warfare in more detail. Let us limit ourselves to two specific elaborations of the concepts, each of which touches upon the basic rationale of avoiding friendly and civilian casualties that is at the heart of proposals for 'Third Wave War.'

A New Look on Interdiction

The first concerns the conceptual changes pursued by the US Army over the 1980s. According to the Toffler's (1994, p 7-10, 51-68) account, the revision of the Army's Field Manual 100-5 and related concepts such as FOFA were significantly inspired by their view on Third Wave economies.[4] They tell us that they have written *War and Anti-War* in close discussion with the officers involved in developing the new Army doctrine.

The couple considers the Second Gulf War as the first serious try-out of the concept, and claim it has been successful. While there also have been second wave mass killings, there have been important elements of third wave war as well. The much televised surgical attacks are an example: the de-massification of war through precision strikes. (Note, in passing, that this is similar to Meilinger's seventh proposition: 'Precision air weapons have redefined the meaning of mass').

This 'new look' on interdiction, an element of third wave war on which the Tofflers are particularly enthusiastic, has found support and elaboration in airpower literature. Meilinger himself, a few years before he published his propositions, has proposed to redefine the notion of 'interdiction' in line with such a view, implying that interdiction focuses on moving targets (Meilinger 1993). Similar views were expressed in the same issue of the *Airpower Journal* by Lt. Gen. (USAF) Buster Glosson, who proposed more reliance on precision guided weapons against mobile targets as they "reduce the human cost of war," and thus

---

[4] This account, to whatever extent it may be accurate, is an interesting and enlightening supplement to the 'official' version of the story as written by Romjue (1984).

"[agree] with American values [, especially the country's] keen intolerance for casualties." (Glosson 1993, p 5,7). More recently, Colonel Bingham (1996) has proposed to "revolutionise warfare through interdiction," against armoured vehicles on the move, thus killing "not many enemy soldiers," "greatly reduc[ing] the risk to civilian lives and infrastructure," and exploiting fear to "achieve success without inflicting immense physical destruction and loss of life." The latter reminds us of another one of Meilinger's propositions, the fifth: airpower can be used to produce physical and psychological shock. All these authors conclude from the interdiction opportunities they discuss that there will be a major role for airpower in future wars. So airpower is again supposed to meet the classic expectations of speed, range and payload. More concretely, it appeals to at least three of Meilinger's ten propositions: (3) airpower is primarily an offensive weapon; (5) airpower can inflict physical and psychological shock by dominating the time dimension and (9) technology and airpower are synergistically related.

The focus being on interdiction against mobile targets, this concept clearly resembles the FOFA concept during the midst of the 1980s. This should not surprise us given the claimed common roots of the two concepts. The main differences are that the strategic objectives have changed from dealing with numerical superiority on the Central European battlefield without nuclear escalation into winning wars against a variety of opponents throughout the world, with as little as casualties as possible. Yet, on a more abstract level, the similarity is that strategic objectives are to be met rather directly by the exploitation of technological advance. The military intent here too is to seize and maintain the initiative and to keep the force ratios in the battle area manageable. The operational requirements are similar to those of FOFA: to complete the targeting process in a short time; to respond rapidly to changed locations of mobile targets; and to call in attacks either in real time or on very short notice (McCabe 1993, p 5).

Weapons of Mass Protection

Another typical element of third wave war is the idea of casualty reduction through the use of non-lethal and anti-lethal weapons. The rationale[5] is found in an analysis of current and future conflicts. These are no longer waged between nation-states, as Clausewitz still assumed. At least one of the parties in most ongoing conflicts is formed by non-state groups such as liberation movements, irregular forces organised by ethnic groups, terrorist organisations and so on. Reasons for traditional nation-states from the northern hemisphere to intervene in conflicts between such parties or to take up fight against such parties include the desire to re-stabilise regions to safeguard northern interests or international stability, or to prevent (further) violations of international law.

It is this state of affairs that gives rise to the rationale of non-lethal weapons. According to another couple, Janet and Chris Morris, "[t]he ability of the devel-

---

[5] While it remains implicit in the Toffler book (1994, p 163-178), this rationale is articulated by the authors on which the Tofflers largely base themselves: Janet and Chris Morris. See e.g. Morris et al. (1995).

oped world's conflict management bodies to set the agenda – to pre-empt crises with early and decisive diplomatic and unconventional action to mitigate such crises with conventional methods – is demonstrably inadequate" (Morris *et al.* 1995, p 18). They list a range of reasons for this judgement. The most fundamental ones are:

- A given crisis may bear no apparent or direct relation or pose no imminent threat to one's own national security;
- Internal and international consensus for timely action is difficult to achieve because of varying evaluations of the seriousness of the threat.
- The developed world's intolerance of casualties when weighed against the casualty tolerance of the (*! – J.G.*) developing world mitigates against the insertion of ground forces should a consensus for action be developed (Morris *et al.* 1995, p 18).

The Morris' conclude that these factors are the rationale for maximising the possibilities for timely, decisive actions with as little casualties as possible. In terms of means, this implies central roles for airpower and so-called *"weapons of mass protection."* Airpower is emphasised for basically two reasons. It is expected that using air rather than ground forces will in most cases lead to casualty reduction among the military on all sides, as well as among civilians. Second, airpower is able to act timely and decisively. Weapons of mass protection are used to bring about the desired effects at a minimal level of destruction. In sum, "[t]he basic values inherent in airpower – deep penetration, broad reach, precision delivery, early entry – must be augmented with the ability to carry payloads whose results enforce policy throughout the operational continuum in ways suitable to the needs of decision makers in the age of chaos" (Morris *et al.* 1995, p 28).

While the Morris' proposal definitely is an attempt to have a more political debate on non-lethal weapons, and, culturally biased and simplifying as it is, makes much less a caricature of the developing world than some other fashionable analyses, it is still based on similar basic assumptions. It is an attempt to meet generic strategic objectives through concepts which are largely based on technologically determined expectations concerning both airpower and non-lethal weapons: technological capabilities are emphasised, and operational limitations and strategic particularities are de-emphasised.

4   Approaches to a Critical Assessment of 'Classical' Visions

If current concepts for 21st century warfare just reflect a vision that in terms of its fundamental assumptions resembles earlier visions, then the same methods that have been used to critically assess these visions with an eye to their fundamental assumptions may be helpful to investigate existing concept for 21st century warfare. An assessment to test the assumption that technological capabilities may enable attainment of strategic objectives, without much consideration of operational limitations, should obviously consider the application of technologies *in their operational context*. This is exactly what critical analyses of established defence planning did, especially during the 1970s and 1980s, when the limits of established thinking were becoming clear. Such assessments typically showed how operational limitations, such as human capabilities, procedural constraints, weather and terrain conditions as well as adversary countermeasures reduced actual combat capabilities to considerably less than technological capabilities would suggest. They also commented upon the strategic merits of established defence planning. To get a sense of the type of technology assessment needed for a fundamental appraisal of proposals for 21st century airpower, let me now briefly discuss the way in which engine technologies, aircraft design complexity and the FOFA concept were critically assessed during the 1970s and 1980s.

Engine Technology

We have already seen how a continuous improvement of engine thrust, thrust-to-weight ratio and fuel efficiency have been part of the 20th century airpower paradigm. These improvements have tremendously changed the face of airpower. Even stronger, especially during the first decades since 1903, these developments have made the air force a viable military service in the first place. It may also be true that until one or two decades ago, technological limitations were still more crucial than operational limitations.

But around 1980, an increasing number of analysts (e.g. Canby 1975; Binkin 1986) started to claim that the point had definitely been reached that operational circumstances are most determining for combat power. Not only had pilot capacities become the major limiting factor, especially in fighters - an important indicator of the wide recognition of this problem being a proliferation of mitigating measures, including training, ergonomics and attempts to develop drugs to increase resistance against extreme accelerations. Also increasing costs of pilot training have led to a situation, in which available airpower and its survivability are to a significant extent determined by the available amount of well-trained manpower (Binkin 1986).

Finally, a criticism was that the development trends just sketched also contributed to tremendous cost increases, that have led to decreased survivability in several ways. But that story can only be fully grasped if we consider another dominant trend in aircraft design: the steadily increasing overall complexity of military aircraft.

Increasing Complexity

Increasing complexity too became subject of scrutiny of an increasing amount of critical analyses (e.g. Spinney 1980; Kaldor 1982; Grin 1990; Demchak 1991; Grin 1992). The claim was that, by then, the trend towards ever increasing complexity has gone beyond the limit of what still can be considered as useful, in more than one respect. First, increasing complexity was said to lead to increasing costs.

Although the costs per function have decreased, this appeared more than compensated by the increase in the number of functions (Walker 1987, p 33-34; Rhea 1990). Moreover, they pointed out that the costs of developing the software required for all these functions and for overall integration are enormous (Walker 1987, p 33-34; Rhea 1990), especially if different functions are interrelated, as is often the case (Demchak 1991, p 27-30). Finally, it was argued, system integration to balance all parts had become very delicate. As a consequence, even relatively small changes in military requirements, technological specifications or financial boundary conditions might imply very elaborate and expensive revisions of the aircraft design and/or component development (Biass 1990, p 19), driving R&D costs which nowadays form a very large share of total aircraft costs (compare Enserink *et al.* 1992, on the EFA).

These cost increases, critics argued, may lead to a variety of disturbing problems (Spinney 1980; Field 1985). One is that increased unit costs in the course of the 1970s started to lead to procurement of relatively low numbers of systems (aptly called 'structural disarmament' by Field 1985). These reduced numbers of available systems negatively impacted on the structural survivability of the air forces. On a higher level of analysis, the situation was even worse: since virtually all major weapon systems were getting more complex, procurement budgets appeared to have increased at the cost of operating budgets. This further increases structural vulnerability. Finally, cost increases led to negative scale effects, as the demand from both the domestic and export markets will decrease.

Another problem attributed to increasing complexity wass that maintenance got more time-consuming and specialised. Table 2.1 portrays the relation between maintenance time and system complexity as illustrated by Kaldor (1982, p 131-132) and by Binkin (1987, p 51), both citing a 1980 Pentagon study as data source. Maybe the most striking fact that emerges from this table is that within a particular aircraft type such as the F-4 Phantom or the F-111, maintenance is most time-consuming for the most complex modification (the F-4J and the F-111D, respectively).

At first sight, the most interesting exception to the rule is the F-15, which, although highly complex in design, requires relatively little maintenance per sortie. This system was designed to simplify maintenance through a modular design that facilitates a 'take-out-and-replace' approach to maintenance; for example, its avionics is concentrated in 45 'black boxes', and plugging rather than soldering is applied at the intermediate level. Moreover, the aircraft has built-in test equipment to monitor these avionics boxes. Faulty boxes are taken out and sent to an avionics intermediate shop (AIS) where a technician identifies the defective card through very sophisticated automated test facilities, and then replaces it.

However, the analyses just mentioned stressed that these AISs are very complex facilities themselves, containing twice as many electronic parts as the F-15 and requiring much maintenance. Moreover, in practice there appear to be AIS supply problems; there are important shortages of parts, including 'black boxes' (compare the above remarks on the more general spare part problem caused by high weapon system costs), and there is a lack of skilled engineers (Binkin 1986, p 55-57, 60-61; Kaldor 1982, p 131). In fact, the number of maintenance personnel per aircraft in a typical F-15 unit is twenty-four, just as much as in a typical F-4 unit.[6]

A variety of consequences were attributed to increased maintenance burdens. First, it makes aircraft more dependent upon specialised bases with all required maintenance facilities and personnel. This is not only costly and, ironically, more a contribution to than a solution for the *problematique* of structural disarmament. Equally ironical is that it may reduce one of the key attributes of airpower, its flexibility. Moreover, it may increase vulnerability because aircraft will stay on their basis (that may be attacked) for a relatively long time as well as because it gets more difficult to use dispersed bases.

Table 2.1 Reliability and maintenance time for aircraft of varying complexity (Source: A 1980 Pentagon study, cited by Kaldor (1982, p 131-132) and Binkin (1987, p 51)).

Aircraft	Complexity	Reliability (Mean time between failures in flight hours)	Maintenance
A-10	Low	1.2	18
A-4M	Low	0.7	29
AV-8A	Low	0.4	44
A-7D	Medium	0.9	24
F-4E	Medium	0.4	38
F-4J	Medium, but higher than F-4E	0.3	83
F-15	High	0.5	34
F-14A	High	0.3	98
F-111F	High	0.3	75
F-111D	High (but higher than F-version: more avionics)	0.2	98
A-6E	High	0.3	71

FOFA

The FOFA concept has been comprehensively assessed in a study by the late US Congress' Office of Technology Assessment (OTA 1987). I will limit myself here to an assessment of a key system, the Joint Surveillance and Target Acquisition System (JSTARS) that was intended to fill the command and control gap implied by the new mission of interdicting *moving* targets.[7] The technological capabilities of this $ 300 million radar aircraft are extremely impressive. Based on a huge F-108 airframe, it can operate at 10-12 km altitude for about 10-11 hours. Its radar

---

[6] It is interesting to note that similar conclusions have been obtained for the another very complex system, designed for modular maintenance, the M-1 tank - compare Demchak (1991, p 97-99, 106, 111-113).

[7] The following data have been gathered from a variety of sources, integrated in the case study in Grin (1990, p 163-182).

has a range of between 180 and 280 km, and is able to detect and track moving targets. Its claimed advantages include a high revisit rate, its large coverage area and depth, its superior capability against moving targets and its 'integrated reconnaissance/attack control capability.' The combination of these features should enable NATO commanders to go through their 'decision and targeting-loop' quicker than their WTO counterparts, and thus contributes to seizing and maintaining the initiative – as we have seen, the very rationale of FOFA as an operational concept to deal with the strategic problem implied by the perceived numerical inferiority *viz-a-viz* the Warsaw Treaty Organisation in a time when it was obvious that resources were limited and public support for nuclear weapons was deteriorating. Given that an 'expensive' solution was not possible and a classical 'cheap' response not acceptable, FOFA entailed the promise of a 'clever (= technologically advanced)' response.

Yet, in a detailed analysis it was argued (Grin 1990) that these impressing technological implications are easily off-set taking into account operational limitations and adversary countermeasures. Because of its sheer size, the aircraft is vulnerable and needs to fly at a stand-off range of 150 km. This fact alone reduces its effective depth against follow-on on forces to 30-130 km. Due to terrain and foliage, its capacity to detect targets declines rapidly beyond a range of 200 km from the platform. Moreover, signals from targets at such ranges will be rather weak and therefore easily disturbed by jammers. In other words, against such a sophisticated systems, countermeasures are relatively easy, just *because* sophistication is inherently vulnerable; and terrain is limiting *because* technological capabilities are so overwhelming. Yet, given that FOFA was based on the idea of technology acting as a force multiplier, we should not be surprised to find severe consequences for the success of the concept.

Indeed, it was shown that even JSTARS has limited capacity for interdiction within the corps level area of influence (which, in Central Europe, was considered to be 80-120 km beyond the forward edge of the battle area). Moreover, in its target attach mode, much of its accuracy would be lost: the delays involved in interpreting data, taking and communicating decisions, preparing missiles and missile flight time would add up to 8-13 minutes, enough for a battalion to move several kilometres...

Other operational circumstances appeared prohibitive for realising its supposed "integrated reconnaissance&attack control capability". In particular, such an integrated capability would require that all just mentioned functions would be performed aboard Joint STARS. Given the enormous complexity of missile units - sophisticated missiles units might have as much as 300 men to operate 8-9 launchers – it is unlikely that an airborne operator would manage to keep sufficiently informed about the status of his missile unit.

In the light of FOFA's strategic ambitions, these conclusions are rather ironical. They imply that it is rather doubtful that the concept enables technology to play a 'force multiplier' role - not only because technology still has to be applied by the forces it is to multiply, but also because it is so complex that it is an extension of, rather than a solution to, the problem of structural disarmament.

5   Assessing 'New' Concepts for 21st Century Airpower

From the previous sections we can deduce what sort of approach to TA is needed to test the validity of the sort of claims associated with the classical horizon of expectations concerning airpower: a contextual consideration of the way in which technology would play its role in the envisaged military posture, taking into account

- not only the technological but also the human and operational elements of the technological system,
- as well as environmental factors such as terrain an adversary countermeasures,
- against the background of the strategic *problematique* involved: it was only against the strategic objective of overcoming numerical inferiority in the face of resource constraints that conclusions concerning the implications of increasing complexity got their full meaning.

Let us now attempt to identify the issues to be covered in critical assessments of Tofflerian concepts for 21st century air warfare, focusing on the 'new look on interdiction' and the idea of 'weapons of mass protection.'

A New Look on Interdiction

In the light of FOFA, one wonders how new the 'new look' on interdiction actually is. Probably the Tofflers are more accurate here when they claim common roots. In any case, and hardly surprising given that FOFA implementation started only during the final years of the Cold War, the same systems designed for FOFA play a role in current concepts for deep interdiction. Is there reason to believe that the operational limitations that were likely to hinder such operations in the FOFA case, do not apply here? According to the proponents of this new look on interdiction, it will be a highly successful way of dealing with the adversary. For instance, Toffler (1994, p 65, 87-88) and Bingham (1996) both sing the praises of Joint STARS because of its success during the second Gulf War.

A more critical type of assessment, of the kind just cited, would emphasise that these claims are based on generalisations from the Second Gulf War, and neglect the particular contextual background of this success. Specifically, such claims appear to neglect the peculiarities of the situation in Iraq: extremely flat terrain, virtually absolute air superiority and thus no countermeasures of any significance, and no intensive, close battle between the respective land forces. It is interesting to note that even Bingham mentions the fact that in those areas where there was difficult terrain and foliage, Joint STARS had difficulties in detecting and tracking targets. Imagine what it might have meant for Joint STARS effectiveness if one or, worse, several of the following circumstances would have been there (compare the analysis in Grin 1990 and McCabe 1993):

- terrain and foliage would have been more adverse in more areas,

- friends and foes would be moving on the ground close to each other, complicating identification and introducing time delays between detection and attack,
- there would have been, at least locally, air superiority on the other side,
- there would have been active electronic countermeasures, such as jamming.

Indeed, during NATO's bombing campaign in Kosovo, following the ethnic cleansing in Kosovo early 1999, terrain, foliage as well as bad weather have seriously hampered air operations. Also time delays between detection and attack have been reported. Limited vision has sometimes caused that strikes had to be cancelled. In other cases, outdated data or limited vision caused refugees convoys or buses with civilians to be hit instead of Serbian tank units (Cook 1999; Bender 1999).

Another aspect neglected by optimistic reports on Joint STARS roles in the second Gulf War concerns the operating bases for Joint STARS. These need enable landing and take-off of heavy aircraft, and contain specialised assets including personnel, equipment and spare parts to repair aircraft. During the Second Gulf War, there was the fortunate circumstance of sufficient preparation time, and a friendly country nearby where bases could be established for Joint STARS. That may well be different however, at conflicts elsewhere around the globe. Moreover, such bases form lucrative targets for offensive counter air attacks or insurgency operations. It is of significant interest to refer here to our earlier discussion of the relation between complexity and maintainability (compare McCabe 1993, p 7). That analysis suggests that these problems will also play a role for other sophisticated aircraft, including those who are to perform deep interdiction sorties to deliver precision guided munitions.

In sum, it is difficult to disagree with Major McCabe (1993, p 6), in one of the few critical analyses of the concept, when he concludes: "When evaluating Desert Storm, we should recognize that US effectiveness was enhanced by Iraq's attempt to wage a positional war rather than a war of maneuver. It remains to be seen if the speed and robustness of the deep attack process can be increased to handle a war where both sides are attempting to wage a war of maneuver and where our command, communications and airpower are central targets of enemy efforts. (...) potential enemies can be expected to (...) seek to devise workable countermeasures."

Weapons of Mass Protection

The idea of 'weapons of mass destruction' is perhaps most vulnerable to a mechanism well known from military history: that adversaries follow a 'different code,' as Van Creveld has put it (van Creveld 1989, p 295-296). One simple, yet effective, example was the Serbian refusal to use their major air defence assets during the early days of NATO bombings in the Kosovo campaign. As a consequence, NATO had to fly high during the rest of the war, in some cases leading to additional casualties.

More macabre breaks of the code are conceivable, of course. In thinking about the contingencies discussed by the Toffler and Morris couples, we do well, for instance, to remember what happened in Bosnia during the Spring of 1995. At that time, after some brutal attacks by Serbs against civilians, the multinational forces in Bosnia attempted to force the Serbs on their knees through disrupting by air attacks their air defence, logistics and command and control assets. Admittedly, these attacks had some success, but they could hardly be continued for a suffi-ciently long time: the Serbs, – operating according to a different code – took UN soldiers as hostages and placed them at remaining assets so as to prevent further attacks. Also, they were on the edge of using another weapon: accusing UN troops of civilian casualties as a consequence of attacking key military assets located in population centres.

If it is true that also potential aggressors' generals prepare themselves for the last war, they may plan in future cases to take recourse to either of these 'weap-ons' against quick disruption. In fact, although still unconfirmed at the time of writing (right at the end of the bombing campaign), there have been persistent reports that the Serbs had captured several thousands of Kosovar Albanians for such purposes. Adequate anticipation of such countermeasures is complicated by the fact that they may differ from case to case, a fact certainly decreasing in im-portance if the Tofflers are right that we will increasingly face 'a collision of war forms.'

Furthermore, and analogously, if northern nation states are going to rely on air-power for such missions, little doubt is possible that potential aggressors will start to devote significantly more resources to advanced air defence. That could reduce both rapidity and decisiveness of these operations. In addition, it may put limits on the possibilities for virtually absolute casualty avoidance In Kosovo, for instance, the choice to bomb from relatively high altitude was inspired by the desire to save airmen's lives, but, as we just noted, it has increased the number of casualties amongst civilians. Of course, aggressors will know their adversary to be vulner-able to TV pictures of casualties flown back to their home countries and to pro-longed battle. Also, if one reads some descriptions of the use and effects of non-lethal weapons, it must really be wondered whether this is the way to ensure better public support - here again, media coverage can easily lead to a completely differ-ent interpretation among the public.

6   TA's for Constructing more Robust Security Policies

This paper started with a reference to the so-called Revolution in Political and Military Affairs. The Revolution appeared to amount to concepts for 'third wave war,' with the objective of shaping technology and operations for multilateral peacekeeping or unilateral intervention contingencies against 'new' adversaries in the face of the challenge of maintaining public support. We have seen that such concepts reflect the classical airpower horizon of expectation: in order to realise strategic objectives, they rely directly and onesidedly on technological capabili-ties, neglecting or at least de-emphasising operational limitations. These concepts emphasise the exploitation of technological progress to do the same, yet better, than any conceivable adversary.

The findings of the above, brief and superficial, scrutiny of new concepts using this type of assessment suggest that the central claims of these concepts may, depending on the operational context, be more or less valid. Would more full-fledged assessments along such lines be undertaken, they would probably form a relevant contribution to the ongoing political judgement about the strategic issue of how to devise a democratically legitimate defence planning that fits the post-Cold War mission of the armed forces. However, as a contribution to political judgement they are also limited, especially in the following respects.

First, what has been shown in the preceding analysis is that it is plausible that the claimed merits of Third Wave War concepts may not apply in all contexts. The other side of the coin is that the validity of analyses like those suggested above is limited to the context of analysis. Even if we would be able to enhance the plausibility through performing a more elaborate assessment, it remains true that the conclusions are context-dependent. This implies two limitations. The first is as simple as troublesome: how can we learn from an analysis for one context with regard to another one? Second, while it may be true that the effectiveness of strategic concepts may be context dependent, it is obvious that defence policy, especially material development and acquisition and personnel training are necessarily to prepare the forces for a variety of contexts. Also, the actual use of military force should be legitimate in all cases.

The third limitation of the type of assessment discussed above is that it yields negative conclusions only: it indicates the shortcomings and Achilles' heels of existing concept. While such critical scrutiny has, of course, a right of its own, TA's relevance to political judgement increases when it can also help to identify, positively, alternative ways of dealing with the strategic problem which third wave war concepts were to solve. A good starting point is to consider the second and third limitations first. How to use contextual analysis for constructing a security policy concept that may be both legitimate and effective in a variety of contexts, yet does not rely on unwarranted, generic claims? The answer is to use such analysis to identify what elements, and how, are sensitive to contextual factors, and then to attempt to design overall policy such that it is able to transcend these limitations. In fact, the preceding sections suggest that we need to do so on two levels: first, regarding the means through which strategic objectives may be attained; and, second, regarding the military means we need to re-appreciate the relation between technology, military concept and operational limitations.

Exactly to be able to benefit from critical assessment in order to construct new, more viable visions, we do well to understand the fundamental roots of the assumptions that we criticise. Keeping a long and complicated story short and simple, we might observe that the types of airpower visions we have encountered are part of the wider programme of High Modernity. In addition to the reliance on technological progress *per se* as a solution to social problems and its elaboration into more specific guiding principles, there are at least two other key elements to that Paradigm. First, there is the view that wise policies can be based on universal strategies, based on an equally universal understanding of the world; and second, the idea that the world order must be based on a stable system of internally and externally sovereign nation states (Toulmin 1991, p 201) has nicely captured the interwovenness of these three elements (reliance on technological progress; the idea of complete, generically valid knowledge; and a world order constituted by sovereign nation states) of the Modern Paradigm. As he puts it, "the seduction of High Modernity lay in its abstract neatness and theoretical simplicity; both of these features blinded the successors of Descartes to the unavoidable complexities of concrete human experience".

Our analysis in section 5 has examined the concepts in terms of two assumptions that reflect this 'seduction'. By taking into account operational circumstances in assessing the validity of claims about the role attributed to technology in airpower visions we have examined the validity of claims concerning the first key element is pulling into question the validity of the second key element: the possibility of acting on the basis of universally certain knowledge. Thus, we have used the epistemological and methodological alternatives to the modern paradigm in order to critically scrutinise typically modern visions.

Let us now use the more substantive elements of the alternative paradigm to sketch the contours of a truly alternative security policy. Specifically, let us for the moment assume that, in the next century, nation states as the central actors at the global stage will increasingly have to share their place with transnational actors such as corporations and non-governmental organisations (NGO's). Closely related, *influence* rather than *power* "will be the name of the game" (Toulmin 1991, p 208). In that game, NGO's may often turn to be primary bearers of moral authority.

Thus, applied to our subject: while military intervention for hegemonistic or purely economical motives may be controversial, humanitarian intervention may attain wide support if NGO's emphasise the fate of an aggressor's victims and point out that they deserve international community's solidarity. Empirically, there is evidence for that viewpoint in surveys reported by van der Meulen (1997). His conclusion is that, in countries as diverse as the US, Spain, France and the Netherlands, it would be wrong to believe that public support is deteriorating *per se*. Much of public judgements depends upon the modalities of the mission under scrutiny: is it undertaken within a multinational (UN) framework, are its objectives unambiguously in line with the proclaimed legitimisation, and so on. The acceptability of the risk of casualties is considered contextually, taking into account such modalities. The most widely shared public doubts on NATO's actions in the Kosovo crisis, concerned the lack of legitimacy in terms of such criteria: the fact that ethnic cleaning was not stopped by the bombing campaign, and the fact that they had been undertaken without UN Security Council involvement. Shaping security policies as a whole, not just their military component, such that they are able to meet criteria like these may also be the best conceivable hedge against adversaries 'fighting according to a different code'. A self-imposed dependence of public legitimacy on casualty avoidance will increase the effectiveness, and thus also the likelihood, of such malign behaviour as hostage taking. Conversely, a publicly expressed determination to fulfil a mission that is widely recognised as legitimate, may reduce vulnerability to such contingencies.

The second question, then, is how to shape airpower in such a way that it may, in a particular context, legitimately and effectively contribute to such a 'package of influence.' That is, we need military concepts that i) exploit technology-in-its-context such that strategic objectives are attained and ii) express the willingness to reduce casualties under *all* parties, but iii) do not make legitimacy solely dependent on casualty minimisation. To prevent that technological capabilities are offset by operational limitations, we need concepts that pay attention to both. Either we need technologies that are less hampered by operational circumstances; or we need military concepts that do not assume that operational circumstances fit maximum exploitation of technological capabilities but, contrarily, seek to take into account or even define operational circumstances. To explore these types of solution, we may benefit from contextual analyses that show us *what* types, precisely, of operational circumstances hamper *what* types of technologies to serve *what* types of military concept. It is precisely in that sense that we *are* able to learn from one context lessons that apply to another one, meeting the first limitation we cited of our contextual analysis. For instance, the Kosovo experience suggests that (Bender 1999):

- laser guided anti-runway bombs are hindered considerably by bad weather, while satellite guided equivalents are much less sensitive to such circumstances.
- the accuracy of satellite guided bombs is, however, sensitive to time delays between target detection and targeting, a problem that increases with the depth of attack. This suggests that satellite-guided munitions may be preferred over laser-guided ones, and that a 'shallow look' on interdiction may be more appropriate than current concepts' 'new look on interdiction.'

On the level of conceptual guiding principles, the above suggests that we reconsider the good old principle 'be the same as the adversary, but do it better.' If we attempt to find a generic solution fitting that principle, there is only one: be better than all conceivable military powers that you may have to fight. This is of course exactly what the visions we have cited are attempting by relying on the idea of technology as way to overcome all operational limitations. We have seen, however, that it is doubtful at best that this can be successful in all cases. It is possible, however, to adopt another adagium: make the adversary operate in a way that fits you – 'let him be worse in the operations that you compel him into.'

To illustrate the notion, let us focus on one key term in existing visions: initiative. In these visions, 'having and maintaining the initiative' is thought to necessarily imply that one should be quicker than the adversary. A much more fruitful approach, however, may be to 'make the other dance to one's own tune,' that is to impose operational constraints on the other side (Grin and Unterseher 1990). An additional motto would be: optimise operational concepts to take account of a variety of operational limitations (terrain conditions, weather circumstances, countermeasures and so on). In practical terms, this would means (e.g. McCabe 1993; Hewish 1995; Unterseher 1994), in terms of operational concepts:

- *Less emphasis on deep interdiction*, since the target acquisition demands will in many cases be too tight to meet. Instead, ensure air superiority at those places where the adversary has no choice but to concentrate his ground forces, that is at moderate distance from the site to be protected, such as a safe haven.
- *More emphasis on passive air defence*, forcing the adversary to disperse his fleet so as to make air defence of friendly troops and civilian protégés easier.
- *Ensure sufficiently high numbers of aircraft*, since numbers remain important to realise superiority, and because we have seen that it is dangerous to rely on high tech *per se* as a force multiplier. In terms of technology, this implies less emphasis om complexity and more on cost control. For the particular case of target acquisition, attack control and attacks itself are involved, the idea of shallow interdiction is indeed well compatible with making use of smaller, less complex aircraft as well as unmanned aerial vehicles (Grin 1990, pp 182-191).

The preceding discussion suggests some elements of a vision that radically departs from typically modern visions. We have now seen how a combination of a critical assessment of modern visions with a deeper understanding of their fundamental assumptions helps us to identify such elements. Yet, while this may be a necessary, it is not a sufficient condition for formulating a more balanced vision. As I emphasised in chapter 1, in order to escape the trap of subsuming one vision under the other, we need some pluriform process of comparatively assessing options that are rooted in different normative perspectives. Thus, while this chapter may have suggested some fruitful approaches critical, comparative analysis, it also makes clear that we may, at present, be short of one crucial prerequisite: the availability of expertise on other than merely modern strategic choices and their elaborations in guiding principles in technological and conceptual development. We do wise to take into account, more than we have done hitherto, the need to maintain, in addition to defence establishments, a range of other think tanks, including institutes that work in close cooperation with NGO's.

References

Albrecht, U (1974) Das Ende des MRCA? In: Studiengruppe Militärpolitik (ed) Ein Anti-Weissbuch, Materialien für eine Alternative Militärpolitik, Rowohlt, Hamburg.

Bender, B (1999) Allies still lack real-time targeting. Janes Defence Weekly. (posted April 7, 1999).

Biass EH (1990) Tomorrow's Fighters: Aircraft or Integrated Weapon Systems? Technology is no longer the number one problem, Armada International, 4/1990, p 18-33.

Bingham PT (1996) Revolutionizing Warfare through interdiction, Airpower Journal, Spring 1996, p 29-35

Binkin M (1986) Military Technology and Defense Manpower. Washington D.C.: The Brookings Institution

Bunker RJ (1996) Generations, Waves and Epochs. Modes of Warfare and the RPMA, Airpower Journal, Spring 1996, p 18-28

Bunker RJ (1997) Technology in a neo-Clausewitzean Setting, in: Gert de Nooy (ed), The Clausewitzean Dictum and the Future of Western Military Strategy, The Hague etc.: Kluwer Law International

Canby S (1975) The Alliance and Europe: Part IV. Military Doctrine and Technology. Adelphi Paper, No. 109. London: The International Institute for Strategic Studies.

Cook N (1999) NATO battles against the elements. Jane's Defence Weekly. Text obtained from www.janes.com/defence/features/kosovo (posted April 20, 1999).

De Jouvenel B (1963) The pure theory of politics. Cambridge: Cambridge University Press

Demchak CC (1991) Military Organizations, Complex Machines, Modernization in the US Armed Services. Ithaca & London: University of Cornell Press

Enserink B (1993) Influencing Military Technological Innovation: Socio-technical networks and the development of the supersonic bomber. Delft: Eburon

Enserink B, Smit WA, Elzen B (1992) Directing a cacophony: weapon technology and international security, in: Wim A. Smit, John Grin, Lev Voronkov (eds), Military technological innovation and stability in a changing world. Amsterdam: VU University Press

Field, B (1985) Economics and defence resources: the prospect. NATO Review, 5/1985, pp 24-29.

Glosson BC (1993) Impact of Precision Guided weapons on air combat operations, Airpower Journal, Summer 1993, p 4-10

Grin J (1990) Military-technological choices and political implications. Command and Control in established NATO posture and a non-provocative defence. Amsterdam/New York: VU University Press/St. Martin's Press

Grin J (1992) Assessing military aircraft innovation and R&D paths, in: Wim A. Smit, John Grin, Lev Voronkov (eds), Military technological innovation and stability in a changing world. VU University Press, p 215-239

Grin J, Unterseher L (1990) ...den Bedrohungszirkel unterbrechen: Spinnennetz. Ein Militärtheoretischer Beitrag zur Um- und Abrüstung, in: Wolfgang Vogt (Hrsg.), Mut zum Frieden. Über die Möglichkeiten einer Friedensentwicklung für das Jahr 2000, Darmstadt: Wissenschaftliche Buchgesellschaft, p 243-262

Grin J, van de Graaf H, Hoppe R (1997) Technology Assessment through Interaction. A Guide. Den Haag: SDU (Working Document Rathenau Institute, W57)

Gunston B (1978) Aviation. The story of flight. London: Hennerwood Publications

Hartcup G (1993) The Silent Revolution. Development of conventional Weapons, 1945-85. London: Brassey's Defence Publishers

Hewish M (1995) Airborne ground surveilance. New technologies meet urgent demands. International Defence review, 1/1995, pp34-39.

Kaldor M (1982) The baroque arsenal. London: Sphere Books

Law J, Callon M (1992) The life and death of an aircraft: a network analysis of technical change, in: Bijker WE and Law J (eds) Shaping technology/building society. Studies in sociotechnical change. Cambridge, MA & London: The MIT Press, p 21-52.

Mason RA(1987) Air Power. An Overview of Roles. London: Brassey's Defence Publishers

McCabe TR (1993) The limits of deep attack, Airpower Journal, Spring, 1993, p 4-14.

Meilinger PS (1993) Towards a new airpower lexicon. Or interdiction: an idea whose time has finally gone? Airpower Journal, Summer 1993, p 39-47

Meilinger PS (1995) Ten propositions regarding air power, Airpower Journal, Spring 1995, p 50; 52-72

Morris C, Morris J, Baines T (1995) Weapons of Mass Protection, Airpower Journal, Spring 1995, p 15-29

Morrow JH (1996) Expectation and Reality. The Great War in the Air, Airpower Journal, Winter 1996, p 27-34

Office of Technology Assessment of the US Congress (OTA 1988) New Technology for NATO. Implementing Follow-On Forces Attack. Washington D.C.: Congress of the US

Rhea J (1990) The next generation of avionics, Air Force Magazine, January, p 68-72

Romjue, JL (1984). From Active Defense to AirLand Battle: the development of Army Doctrine, 1973-1982. Fort Monroe, VA: TRADOC.

Smit WA (1989) Defence Technology Assessment and the control of emerging technologies, in: Ter Borg, Marlies & Wim A. Smit (eds) Non-provocative defence as a principle for arms reduction and its implications for assessing defence technology. Amsterdam: VU University Press, p 61-76

Spinney FC (1980) Defense Facts of Life. Washington D.C.: unofficial document of the DoD

Szafranski R (1995) Twelve principles emerging from ten propositions, Airpower Journal, Spring 1995, p 51; 73-80

Ter Borg M (1985) Innovatie tot in eeuwigheid. Het geloof in technische vooruitgang in discussie. Amersfoort: De Horstink.

Toffler A, Toffler H (1994; first edition 1993). War and Anti-War. Survival at the dawn of the 21st century. London: Warner Books.

Toulmin S (1991) Cosmopolis. The Hidden Agenda of Modernity. Chicago: The University of Chicago Press

Unterseher L (1994) Air Power and Confidence Building Defense. Research Note. Bonn: Study Group on Alternative Security Policy

van Creveld M (1989) Technology and War. New York: The Free Press

van der Meulen J (1997) Post-modern societies and future support for military missions, in: Gert de Nooy (ed.). The Clausewitzean Dictum and the Future of Western Military Strategy, The Hague etc.: Kluwer Law International, p 59-74

van Lente, D (1988). Techniek en ideologie. Opvattingen over de maatschappelijke betekenis van technische vernieuwingen in Nederland, 1850-1920. Groningen: Wolters-Noordhoff/Forsten.

Walker JR (1987) Air-to-Ground Operations. London: Brassey's Defence Publishers.

III   Visions and Societal Rationality

Technology Policy Between Long-Term Planning Requirements and Short-Ranged Acceptance Problems. New Challenges for Technology Assessment

Armin Grunwald

1 Introduction

Modern societies are facing strong demands for a reliable *long-term orientation* of technology policy, environmental policy and science policy. These demands result, on the one hand, from the very complex nature and the extended time frame of research and development processes and, on the other hand, from the aim to realise the agenda of Sustainable Development (Kuik and Verbruggen 1992). Technology and environmental policy, however, must – in pluralistic and democratic societies – also be based on certain forms of *acceptance*, otherwise its success would be questioned on principle.

> **Box 1:** *To make this dilemma more concrete, consider the present discussion about the greenhouse effect and its presumed consequences for the global climate (global warming). This challenge to technology and environmental policy – the question by what political measures this challenge should be dealt with is still discussed controversially - serves as an example to illustrate the central items of this chapter. A substantial position in this field will not be favoured.[1] Instead, exclusively the structure of the decisions to be made, of the decision-making situation and of its implications will be used in the ongoing text to exemplify the general analyses presented. The field of climate policy (Brauch 1997; Kuik and Verbruggen 1992) includes both the requirements for long-term considerations as well as the problems of the short-ranged acceptance of political steering mechanisms.*

This chapter is devoted to ways of dealing with this dilemma, which seems to be inherent to present, pluralistic and democratic societies in managing technology problems. In spite of the far-reaching nature of the dilemma intimated in the title it has not yet been tackled directly in the conceptual work on Technology Assess-

---

[1] The European Academy is conducting an interdisciplinary project on „Climate Prediction and Precaution" to deal with these questions from a very general point of view to improve the rationality of the debate („rationality" used in the pragmatic sense outlined in section 5, *not* in order to oppose the rationality of experts and the „irrationality" of the general public which is sometimes presumed).

ment (TA). The expectation of the author is that uncovering this dilemma and pointing to a feasible way of dealing with it could help to handle some of the well-known dilemmas in technology policy and TA (Banse and Friedrich 1996) in a more suitable way. More specifically, this issue is of particular importance to the subject of this book, assessing visions. The reasons is, of course, that these visions are characterised by a relatively long time horizon and since they are to guide collective action over a prolonged time.

The approach presented in this chapter claims to show a perspective for resolving the dilemma. It is acknowledged that *without* respect to acceptance at all, reliable long-term planning cannot be reached in a democratic and pluralistic society, and that relying *exclusively* on acceptance does not allow long-term plans to be followed. The way proposed is to *question the concept of acceptance itself and – as a result - to shift the level of acceptance* required. The question to be answered is not what technology will be accepted but *what elements of technology policy must really be based on acceptance.*

The proposal is to focus on criteria for the *acceptability* of technology and on procedures to derive them. The level of acceptance required is, thus, shifted from the acceptance of the factual technology to the acceptance of the rationally justified criteria and procedures to be followed in technology policy – *from substantial acceptance to a procedural one.* These procedures and criteria, on their turn, may be grounded in underling rationality standards of society. Technology policy then can be based on rational (and relatively stable) acceptability criteria and procedures rather than on pure acceptance often depending on chance events.[2] Such an approach is not intended to re-establish past approaches of planning euphoria (discussed in Grunwald 1999a). Instead, an approach to technology policy shall be developed which allows acceptance aspects to be taken into account *without* loosing completely the chance to follow long-term oriented goals.

The path of rationality in this field is, as in most cases (Hartmann and Janich 1996; 1998), the pragmatic midway. It results from performing a careful process of weighing the various arguments involved, instead of simply either relying on acceptance a specific technology as such or on authoritarian dictate. Pragmatic rationality allows even utopian ideas about the future of technology and society to be discussed to orient present-day policy decisions if – and this is of eminent importance – it is *accompanied by permanent reflection.* This reflection should be concentrated on the impact of the ensuing development on society and its members in order to avoid one-sided developments with non-intended side-effects and to uncover and avoid authoritarian aspects of these utopian ideas (compare the critical review on some recent literature on utopian ideas in the introductory chapter by John Grin).

This chapter is structured as follows. In Part I the central dilemma is described and analysed in detail to make the starting point as transparent and clear as possi-

---

[2] The question if and to what extent laypersons should be involved in decision-making processes in technology policy shall not be tackled in this paper. The concept of rationality used here (section 5, Rescher 1988) is dedicated to the power of argumentation independent from whether an expert, a politician or a layperson brings up or supports the arguments. Nevertheless, the approach presented here allows some conclusions for this question (section 9.2).

ble. First, the need for long-term planning is discussed (section 2). Then the orientation of current technology policy on short term acceptance is discussed (section 3). Its shortcomings are indicated, which leads to a formulation of the dilemma on which this chapter focuses (section 4). In order transcend this dilemma, it is useful to refer back, in Part II, to some philosophical ideas on rationality and culture (section 5) as well as on models of technology development (section 6). In Part III, the proposed rational TA is elaborated. The discussion starts from these basic reflections to clarify the concept of acceptability (section 7) and to consider it in terms of its compatibility with democracy (section 8). Finally, the consequences for technology assessment are elaborated (section 9).

This contribution has been developed on the basis of philosophical planning theory (Grunwald 1999a), methodological and conceptual aspects of Technology Assessment (TA) (Grunwald 1994; 1998a; 1999b, 1999c) and the ethics of technology (Grunwald 1996a; 1999d; 1999e, 1999f). It should be understood as a basic programmatic proposal on the role of TA in shaping the future of technology in society by questioning back to some principal problems often hidden behind the ongoing TA debate. The solution proposed, however, does not lead to a simple procedural means directly applicable to any situation. Instead, it is shown that rational technology policy is a difficult endeavour committed to very severe obligations to reflect on the cultural grounds technology is developed upon. Eventually, this leads to the concept of a *learning society* which is reflecting its own cultural and technological evolution relative to underlying rationality criteria to assess this evolution and, vice versa, to develop these rationality criteria beyond their factual present-day relevance to meet the future requirements of technological and scientific advance.

Part I   The Central Dilemma

2   Long-Term Planning Requirements

As is well-known, the economic and technological change in the industrialised countries is accelerating to a considerable extent (e.g. Lübbe 1997). Scientific and technological knowledge becomes antiquated faster and faster. Nevertheless – and this is the main thesis of this section and an important premise of the chapter -, modern societies are facing strong demands for reliable *long-term orientations* of technology policy, science policy and environmental policy. These demands are resulting from quite different sources: (1) from the complex nature and the extending time frame of many processes of research and development as well as of establishing economic innovations and (2) from the requirements of realising Sustainable Development in a globalised world.

(1) There are several reasons for looking for some long-term *reliability* from the point of view of optimising the innovation processes. *At first*, tight financial

and human resources in society require a far-sighted, long-term and rational orientation of scientific and technological developments. Serious course corrections and modifications of major scientific and technological projects at an advanced stage of development often require very considerable resources; an interruption to a line of development leaves corresponding *investment ruins* behind it and can also involve unwelcome political problems as a consequence (Gethmann 1998, Grunwald 1998b). For example, the International Space Station ISS currently being implemented in space is an example for a technology system with large political implications. Its integration time of about ten years shows the necessity of a relatively stable political context in order to ensure finishing and, later on, operating the station. How disastrous neglect of this aspect may be has become clear from the way in which nuclear power technology was pushed with realisation time frames of about five to ten years and lifetimes (and time frames for the expected Return on Investment) of about thirty years. This has left some very expensive investments ruins. For instance, in Germany, the nuclear plants at Kalkar and Mülheim-Kärlich are prominent examples. Importantly, in this context, the main reason for these failed investments might have been the low acceptance in the population.

Secondly, the rapid change of technology will become economically inefficient beyond a certain rate of changing. Innovations (technological ones as well as social ones) always have *destructive* aspects (Schumpeter 1934): new technologies diminish the value of the elder, well-established technologies and lead to a re-evaluation of all related products and services and to transfer costs at several stages of the substitution process. For example, any new software causes adaptation costs because the users have to become familiar with it, e.g. by special training. Innovations will fail economically beyond certain break-even points with respect to the rate of change. The maximum rate of technological change compatible with economic rationality, however, seems to be dependent on the *stability of the surrounding social framework*.

Thirdly, if there were no relatively stable context conditions, allowing calculations and expectations of a Return on Investment, only few would sponsor research and development processes requiring a lot of time and, therefore, bearing many risks of failure. Scientific and technological knowledge must be produced, validated, transferred and applied to technical contexts. Technological innovation can be successful only if there are certain stability conditions fulfilled which allow investment (of time, knowledge and money) into such long-lasting developments (Law and Bijker 1994).[3] As an example, research and development in the materials sciences require, as a rule, from 15 to 20 years to develop new materials in such a way that they can be used in the industrial process (Harig and Langenbach 1999). Therefore, technology policy has to be, at least in parts, designed to the long run in order to ensure *planning security*

---

[3] The contradiction between the notion of an accelerating technological change and the requirements for a relative stability is only a seeming one. Note that the seemingly contradictory observations are concerning different levels of society: the acceleration is going on at the *economic* level, the stability conditions must be guaranteed by the *political* system dependent on the cultural background (section 5.2).

for investments which has been shown to be an important factor for any innovative economy (Hohmeyer et al. 1996).

Fourthly, and this leads to the crux question of the contributions assembled in this book (compare the introductory paper by John Grin), if technology policy is – at least in parts – ruled by expectations about the future and by visions the main thesis of this section seems to be evident because such expectations and visions *cannot be changed rapidly*. Otherwise they would lose their value for social orientation immediately. We cannot abandon our visions and expectations about the future and replace them by some new expectations in a too rapid way because these expectations are related to our culture, our traditions and morals – they are related to our *past* (cf. the introductory chapter to this book by John Grin and section 5 of this chapter). Because we cannot change our past – we can only influence the *assessments* of our past – we cannot change our „futuribles" *in an arbitrary way*. The past in its present-day assessment is the basis for normative expectations about the future, about desirables, purposes and goals (section 7).

(2) To arrive at a full implementation of Sustainable Development – whatever this means in detail – is *per definitionem* a long-term challenge to society (Grunwald 1999b, Kuik and Verbruggen 1992). Sustainable development implies that societal processes of development – including technology development – are re-oriented so as to ensure that the needs of future generations and those of the southern hemisphere are taken into account as well as those of current generations at the northern hemisphere. Thus sustainable development includes necessarily dealing with long term considerations in technology policy. It involves taking into account (normative) aspects of the distant future, of the impact of our present concepts of technology and society on this future and the impact of such reflections on our present-day concepts and ideas (*backcasting*). The time frame of Sustainable Development has to be, for analytic reasons, always the long run.

Decisions, measures, plans or other activities set up to realise Sustainable (Technology) Development must, therefore, include long-term aspects and take into account the presumed expectations and demands of future generations. Proceeding in a purely incremental way does not meet these requirements.[4] Environmental policy towards Sustainable Development has to include some normative and deductive aspects of backcasting, derived from far-ranging expectations. The question is how to reach a common societal understanding of this situation, and, *a fortiori*, a common understanding of what should be done. More specifically, - and this leads to the main topic of the present chapter – the question is in what way it could be achieved *that this common understanding can be made relatively stable and reliable over a certain relevant time frame*. Obviously, it would be counterproductive if the objectives and concepts of sustainability and the measures to be

---

[4] The kernel of incrementalism consists of the approach of trial and error, as suggested in some incrementalistic planning theories (Braybrooke and Lindblom 1963, compare the discussions in Camhis 1979 and Grunwald 1999a). More on incrementalism and technology policy is included in the sections 4 and 6.3.

undertaken to reach these objectives were changing in a rapid way depending on chance events. Though they have to *change* according to new scientific or technological knowledge or according to new and better concepts which may have been worked out in the meantime, the central point is that there must be a *basic understanding* of sustainability which should – at least in "normal cases" - *not change dramatically*. Sustainable Development is only possible with some aspects of continuity and stability.

This does not only apply to the level of rationales but already to the level of political measures. Consider, for example, the envisaged reduction of greenhouse gases set up by the Rio process. Within this process long-term rationales have been chosen (Germany wants to reduce the emission of carbondioxide up to the year 2005 by one quarter compared to 1990). Because the steering mechanisms to arrive at this aim (the eco-taxes, for example) are very complex and often not very well-known with respect to their impact and consequences, it is necessary to gain experience with these measures. Therefore, it is indispensable to spend enough time in applying these mechanisms to society (in the form of a – carefully planned and prepared - large real-time experiment, not in the way of unguided trial and error but on the basis of a careful reflection of those measures and their presumed impact). Applying different measures in a rather rapidly varying way would not even allow an assessment of whether certain measures result in success or failure in order to reach the rationales defined before. If *learning* in such complex fields is allowed and appreciated – and this should be reasonably the case – sufficient time for empirical investigations is required.

> **Box 2:** *For example, all essentials of the global warming example mentioned above are related to the long run: the genesis reaching back to the 19th century's industry, the (still controversial) diagnosis concerning the second or third or even later generations after our generation, and the measures for avoiding an uncontrolled global warming effect.*

In conclusion, the importance of long-term considerations in technology and environmental policy is to be emphasised. For several reasons the time frame of societal actions in these fields has to include some – *principally changeable but relatively stable* – aspects of long-term planning or visions giving orientation how to shape the future of society by technology.

3   Short-Ranged Acceptance-Orientation

Technology policy nowadays is often required to be oriented to the present-day or assumed future level of acceptance towards technology (cf., for example, Todt 1997, Todt and Lujan 1998; Bröchler 1998). The decreasing acceptance of some key technologies (nuclear technology, high speed transportation systems, new airports and gene technology) since the seventies in parts of the population of some European countries has heavily influenced technology policy. Partly the resistance against those technologies is motivated by environmental protection objectives presupposing that those technologies will destroy the natural environment in the long term. On the other hand, there is a growing *acceptance problem of environmental protection itself*. For example, in Germany the resistance against new bio-reservates is increasing in parts of the population affected (like fishing industry, farmers, local industry or even the tourist sector which are afraid to be burdened with economic disadvantages). Here the paradox situation emerges that environmental policy – which has been mainly set up because polluting technologies were leading to severe acceptance problems – runs into acceptance problems itself. The aspect of technology acceptance shall be analysed below in more detail.

Many studies on the acceptance of key technologies or technology in general have been conducted since the early eighties. In some countries monitoring procedures have been established to observe any change of the level of acceptance in the population.[5] Politics and industry are often worrying about the presumed decreasing acceptance of modern technology in some European countries (especially in Germany) because loss of the capability for global competition is feared as a consequence. As a consequence, political decisions relating to science and technology are presently often influenced by estimates of the current degree of acceptance of science and technology among the people concerned. There is, however, an obvious difference between these (often somewhat lamenting) statements and the results of certain polls showing that only some technologies are facing this acceptance problem but not technology in general (TAB 1997). In spite of these empirical results the story of decreasing technology acceptance seemingly does not come to an end and leads to attempts to remove this presumably missing acceptance (see below). Technology policy oriented mainly to such acceptance aspects can be designed in two different ways (Grunwald 1999b):[6]

1. *Adaptive approach*: The experience of technology conflicts since the seventies around some key technologies like nuclear technology or the gene technology (which sometimes led to war-like scenarios in some countries), raised the question whether it would be possible to avoid such conflicts *a priori*. The idea behind this approach is that technology conflicts could be avoided by taking into account the presumed acceptance in technology development and design, i.e. by developing technology in accordance with the values, norms and fears of people.[7]

2. *Shaping approach*: Instead of regarding already existing public acceptance as a rigid determining factor for technology policy making, attempts have been made to influence this level of acceptance – to 'construct' acceptance by information, by increased investments in the 'public understanding of science', in science centres and museums or in certain types of technology-related education or by giving prizes for inventions.

---

[5] For instance, the Bureau of Technology Assessment at the German Parliament (TAB) is conducting such a monitoring activity (TAB 1997).

[6] Technology policy shall be denoted as „acceptance-oriented" within this paper if it is designed to one of these types and, therefore, puts the concept of acceptance into its center of interest.

[7] Compare the discussion on „socially compatible shaping of technology" (*sozialverträgliche Technikgestaltung*) in Germany (Grunwald 1996b and the literature mentioned there).

Within the adaptive approach the *stakeholders* of technology development (customers, citizens, political parties, authorities, social movements - all groups or persons affected by technology policy) are involved in the decision-making process. The degree of involvement ranges from real participation in the decision-making processes to measuring the rates of acceptance by polls. The assumption is that, if the people concerned are involved in the decision-making process, the result should find acceptance among them: „Technologies developed through such strategies will be socially more viable and accepted, which will enhance the economic viability of new products and processes" (Rip et al. 1995b, p 5).

While the first way of dealing with (presumed or real) acceptance problems is a more passive one, by acknowledging the factual rate of acceptance as a strict boundary condition for technology policy, the second way carries an active moment: the acceptance is viewed as a parameter open for shaping. The 'shaping' approach is followed mainly by scientists and technicians (for example, compare VDI 1997). It is confronted with the suspicion to attempt to „persuade" people to accept technology and technology policy decisions from a technocratic point of view. In this view, the reproach is that the 'shaping' approach simply consists of certain forms of advertising declared as education or information.

Such an acceptance-orientation is self-evident and unquestioned in the area of technology development for the free market: technology will of course be designed and developed to reach acceptance otherwise the ongoing investments in new technology would not lead to any return of investment and profit. Acceptance is, in that field, taken into account by market research and demand predictions or it is – „constructively" - to be shaped by means of public relations and advertising. Obviously, these mechanisms show some parallels or analogies of the areas of technology policy mentioned above and the market-oriented economy. One of the main points of this chapter is, however, that there are very severe limitations to this parallellisation according to the very different legitimisation requirements of both fields (section 8). The approaches mentioned above cannot adequately tackle all the various problems involved in technology policy making. Furthermore, the impact caused by over-emphasising the aspect of acceptance might result in unintended consequences (section 4).

4   Shortcomings of an Acceptance-Oriented Technology Policy

It follows from the previous sections, that there is a gap between the requirements for long-term considerations and stability, and the mere acceptance-orientation in technology and environmental policy. Of course, the acceptance of technology and of technology policy is an important factor for managing technology in society (Rip et al. 1995a). No matter how reasonable it may be to consider acceptance as a guiding rule of political strategy, policies relying *exclusively* on acceptance aspects or, at least, *over-estimating* them, are insufficient to deal with the manifold problems involved. In the following some main arguments from risk research, planning theory and ethics supporting this view are assembled, which reveal a deep dilemma in the field of technology policy making (Grunwald 1999b).

4.1
Avoiding Technology Conflicts *a priori* is Impossible

The promised goal of avoiding technology conflicts *a priori* (compare the 'adaptive' approach mentioned in section 3) cannot be ensured by a merely acceptance-oriented technology policy because only *current* acceptance levels can be empirically determined. The rate of acceptance may decrease and technical ventures may, in the worst case, lose acceptance completely, though acceptance was given at the beginning. Predictions of acceptance levels are nearly impossible because of their short-ranged variability (see below, section 4.2).

To handle these problems with an "acceptance-orientation at second order", i.e. developing technology in accordance with the actually given (and varying) rates of acceptance, will be impossible because in most cases large investments (in financial, political or other dimensions) will already have been undertaken which do not allow technology or environmental policy to simply follow the varying rate of acceptance (this seems to be a variant of the well-known dilemma of control, Collingridge 1980, Wagner-Döbler 1989). Such developments *cannot be simply reversed if acceptance is lost*. Policy then must either push technology where acceptance is missing (e.g., the CASTOR transports in Germany) or must handle large investment ruins (like the nuclear power plant at Kalkar). Accordingly, focussing on the acceptance factor neither precludes acceptance problems nor prevents technology conflicts apriori.

4.2
Rapid Changes of the Rate of Acceptance

Acceptance may be very instable in society. Acceptance is related to perceptions of benefits and risks with regard to technological options which depend - as investigations in the field of risk research have shown (for example, Jungermann and Slovic 1993; Renn 1992; also several contributions in Bayerische Rück 1992) - on many factors in the given situation. Such factors include the subjective sensitivity to risks, subjective expectations with regard to usefullness compared with the risks perceived, the temporal or spatial distance from the risks, on whether the hazardeous solution has been freely chosen or enforced, on whether an accident in the area concerned has happened in the recent past etc.

Due to the possibly rapid changes of risk perception due to some contingent events the acceptance of risks may vary, too, resulting in a varying acceptance of the risky technology.[8] If science and technology policy were to be based mainly

---

[8] Consider, for example, accidents in big technologies. Especially the Three Miles Island (1980) and Cernobyl (1986) accidents played a major role for the decreasing acceptance of the nuclear power technology.

on the current rate of acceptance, they would have to follow such contextual conditions as they, possibly, change in a rapid and unpredictable way. Shaping the future based on such unstable conditions would become an incremental process *without direction* (disjointed incrementalism, Braybrooke and Lindblom 1963, Camhis 1979). Continuity, long-term planning security and stability cannot be achieved in this way.[9]

Furthermore, designing technology policy as to meet the factual acceptance might be not very reasonable in all situations. Consider the example of a severe accident decreasing dramatically the acceptance of a transportation system (the Eschede accident of the high speed train in Germany in 1998). Should this transportation system be abandoned as a „lesson learnt" from that accident? Independent of how this question is answered it has to be claimed that the accident itself does not justify abandoning the system because *it does not change the safety or the risk of that transportation system.* If one takes the safety standard as a criterium for decision-making there wouldn't be any need for rejecting that technology because its safety did not change at the occasion of the Eschede accident. Often, however, the opposite case seems to be more plausible, because such accidents give rise to an improvement in the security of the system concerned, with the consequence, that after such accidents the risk might be lower than before.

The acceptance of environmental protection requirements according to the demand for Sustainable Development is dependent on risk perception in some fields, too. Since the "Alarmism" and "Catastrophism" of the eighties have proven wrong, tendencies to understate global environmental problems and problems of developing countries are growing. This is happening due to the pure *perception* that the risks emerging from these fields might decrease in relation to risks coming from other fields, related more closely to the lifeworlds of the industrialised countries (decreasing level of employment, decreasing stability of the social assurance systems in most European countries, for example). But obviously, though the perception of global environmental problems may be decreasing at the present, this does not imply that the risks are decreasing, too.

> **Box 3:** *The predictions of global warming are true, false, more or less founded* independent *from the actual perception and the resulting acceptance of some (perhaps unwelcome) steering activities like eco-taxes. The challenge of global warming is a long-term one, with respect to its genesis, to its diagnosis as well as to the implementation of mitigation strategies. It cannot be tackled adequately by using short-ranged and perception-dependent rates of acceptance as the main driving force.*

The incrementalistic development resulting from a mainly acceptance-oriented technology policy would look like – observed from an external point of view –, indeed, as if "we do stagger through history like a drunk putting one disjointed

---

[9] Lindblom himself has, in later papers, designated this understanding of the incrementalist approach as a misunderstanding. Incrementalism should, instead, be guided by strategic analysis preventing the non-intended impact described above (I am indebted to John Grin for informing me about this background).

incremental foot after another" (Boulding 1964, S. 931) – a really impressive
metaphor for the risks of an acceptance-oriented policy (see fig. 2 in section 6.3).
Such an incrementalism may be the best option for many areas of technology
development (for example, in the industrial processes of designing new versions
of products like software packages). But it is easy to see that the disjointed incrementalism is not suitable for the requirements of a policy directed to Sustainable
Development or other fields with some long-term planning aspects because in
such fields at least the *direction* of certain agendas has to be. to some degree,
stable over time (section 2).

4.3
Anti-Innovative Impact

A focus on actual acceptance is principally anti-innovative in the sense that *incrementalistic* options may be preferred *only* because they presumably will be better
accepted than innovative leaps (Bechmann et al. 1994, Grunwald 1996b). This
means that scientific and technological options may be ignored, if they are
(seemingly) not sufficiently acceptable to the general public *at the time the decision is made.* This approach may have the result that solutions for certain problems which are ideal from a technological, economical, political or ethical point of
view are not included among the set of options before the decision is to be made.
Clear assessments and unbiassed decisions will thus become hindered. Acceptance-orientation often leads to "conservative" policies.

Of course, this chapter does not aim to praise technological innovation indis-
criminately; the „dark side of innovation", its „destructive power" (Schumpeter
1934) which appears in technology development to be a *deep ambivalence* must
be observed and analysed contextually and very carefully (section 9.3 for criteria
of such an assessment).[10] In taking care of this ambivalence the listening to real or
presumed acceptance problems may, and this should be self-evident, result in an
increased sensitivity to possible and hypothetic problems with technology and can,
in this general form, be appreciated. The point to be addressed in this section is,
however, - to make it as clear as possible - that the incrementalistic way of dealing with technology may lead to a rejection of innovative technology options
already in a too early stage of the decision-making process. The incrementalistic
appproach favours decisions which might be *too precautionary* in certain situations. This bias becomes relevant, more precisely, only if the acceptance-oriented
approach is taken in a *decontextualised* form. Obviously, in certain situations it
might be the „best" approach; but probably it does not fit to all situations. At this
point a demand for contextual consideration occurs which relates to the "planning-
theoretical" approach suggested for technology assessment (section 6.2) and

---

[10] See the paper of Rob Reuzel and Gert van der Wilt in this volume where a very impressive
case study is given showing that the ambivalence of the medical technology in discussion
(Cochlear Implantation) was not uncovered by TA but by people concerned (parents of deaf
children and their organisations).

shows the danger of overall approaches with too much emphasis on decontextualisation.

4.4
Naturalistic Fallacy

Relying on factual acceptance in normative questions raises the well-known
problem of the naturalistic fallacy (Moore 1904). Factual acceptance does not
automatically create ethical legitimisation. The Is-Ought problem (Schurz 1995)
consists of a very deep philosophical challenge how to create legitimisation for
normative principles or action requests. It is widely accepted that there is no simple way from the Is to the Ought. If there is a recommendation, an appeal or a
prescription given without normativity included in its premises, one would designate this as „naturalistic fallacy".

This argument is of special importance if expectations and visions shall be assessed. The questions acceptance-oriented technology policy cannot answer are:
Why should those visions be the "best" for which the rate of acceptance is the
highest? Isn't it imaginable to overrule such merely acceptance-oriented decisions? In what way and by what means can legitimisation be achieved (section
7.3)? The (very severe) problem of constituting legitimisation and justifying a
"rational choice" among several options cannot be solved sufficiently by the acceptance-oriented approach alone.

4.5
The Individual Advantage and the "Common Good"

Furthermore, acceptance-oriented technology policy is running into the well-
known problem that there are often contradictory requests from, on the one hand,
the individual advantage of stakeholders and, on the other, the "common good"
(whatever this means in detail). By relying on factual acceptance alone, technology policy would presuppose that the mere aggregation of the beliefs in individual
advantages – which, in fact, dominate in most cases acceptance behaviour – constituted the common good.

It is, however, very easy to give examples where this presupposition can be
shown to be unfulfilled. Often, in political agendas it is impossible to arrive at
Pareto-optimal decisions, i.e. at decisions leading to advantages or, at least, to
non-disadvantages for the persons affected because, as a rule, implementing political decisions will produce winners as well as losers.

> **Box 4:** *Consider the global warming example and the presently discussed instrument of eco-taxes. Eco-taxes would cause short-ranged disadvantages for many people and parts of industry, though these measures are presumably suitable to achieve sustainability in the long run. Consequently, the acceptance of eco-taxes seems to be rather low in spite of the fact that they seem to realise certain aspects of the "common good." The latter, in this case, impliestaking into account the presumed interests of future generations – a real long-term consideration which illustrates very impressively the difference between short-ranged acceptance and the "common good".[11]*

Furthermore, the theory of collective decisions has shown – under very general premises – that it is not possible to aggregate the individual preferences in a reasonable way to get a well-defined common welfare function (the well-known Arrow-Theorem, Arrow 1963, Kern and Nida-Rümelin 1994). Mere acceptance derived from individual preferences does not automatically create the common good but leads to indifferent and ambigous decision-making situations. Taking into account the acceptance behaviour is a valuable step in policy making and belongs to political prudency (section 9); it does not, however, free political authorities to define what should be understood by the common good contextually, in the given situation. It is this difference that we all are facing when following political decisions and juridicial codifications of such decisions even if such decisions are not in accordance with our individual preferences and interests. It leads to the cernel of the present chapter (sections 5, 7).

4.6
The Central Dilemma

Prima facie the best way to avoid technology conflicts could be seen as to design technology and environmental policy so as to meet the presumed acceptance (section 3). As a consequence, the short-ranged flexibility gained according to the varying level of acceptance arises together with a loss of continuity and stability. However, short-ranged acceptance-orientation, summarising the sections above, leads to difficulties in shaping the future because of two effects:

1. The acceptance to be reached can always be only the *present* acceptance; it is not possible to predict the level of acceptance because of its possibly fast-varying nature and its dependence on singular (and unpredictable) events like accidents (section 5.2). Therefore, it is impossible to design technology to meet the *future* level of acceptance (section 5.1).
2. Shaping the future requires some medium- or long-term visions for giving orientation (section 2, compare the case studies by Peter Mambrey and John

---

[11] It seems to be one of the most challenging tasks of philosophical ethics to reflect this gap between individual interests and preferences on one side and the requirements of politically defining what is to be understood in a given situation as the ‚common good' on the other and to suggest solutions for this problem.

Grin in this volume). But if factual acceptance is the only parameter for shaping technology policy, the well-known problem arises that people mostly prefer short-ranged advantages to long-term obligations, burdening them with some more or less uncomfortable confinements to reach some goal in a future relatively far away from their life-worlds (section 4.5).

Thus, technology policy relying only on factual acceptance is merely extending or lengthening the present to the future and not really *shaping* the future by intended and reflected ideas about how we wish to live in future, what technology we wish to develop and to implement and how we wish to face future challenges. Purely incremental processes (perhaps a provocative thesis) *lose the future*, they are *processing only the present-day acceptance level*.

This dilemma is not restricted to technology policy but relates to the general problem of whether and how far the factual acceptance of the results of democratically legitimised procedures is required. The main question derived from this dilemma is how to fulfill the demand for long-term considerations, taking into account the indispensability of an acceptance-orientation at a certain level. How to overcome short-ranged acceptance problems in long-term agendas? The point is that long-term planning needs a kind of *normativity* to be sufficiently robust even in times of minor acceptance. Sustaining such long-term agendas often is a necessary precondition for their success.

But where should this normativity come from? Who (persons or institutions) should be legitimised to decide that low or missing acceptance in certain fields does not prevent the relevant institutions of society from proceeding with the plan under consideration?

Box 5: *With respect to the debate on Sustainable Development the requirements to persist in long-term agendas facing the short-ranged aspects of modern pluralistic societies (like the election period of four or five years) has been discussed under the topic of an "ecodictatory regime" which should, according to the view of some people, ensure the conditions for sustainability also in times of missing acceptance, or even stronger, in times of resistance. This discussion leads in the wrong direction. Undemocratic paternalism – though it would indeed enable long-term planning – is completely inacceptable for ethical reasons.*

The approach proposed in this chapter is, in contrast, to look for rationality standards implicitly inherent in ongoing societal practices and to make these *implicit* standards *explicit* in order to reconstruct the underlying normativity of society. This normativity – if it were possible to proceed as suggested - would not be forced upon society because it would already be accepted *implicitly*. To work out this approach it is necessary to go back to some philosophical ideas about rationality, culture and society.

Part II   Technology Policy and Pragmatic Rationality

5   The Concept of Pragmatic Rationality

The concept of rationality as a legacy of the European Enlightenment has been heavily criticised during the last decades from several standpoints. The "linguistic turn" of philosophy uncovered the role of language in epistemology. Critical Theory rejected the "technically halved rationality" (Habermas) and the "one-dimensional Man" (Marcuse) as reductionist programmes. Critical Rationalism rejected the claim for verifying scientific theories. Finally, perhaps the most radical criticism came from postmodernity which questioned the concept of rationality principally, dissolving the concept of a unique rationality postulated by the European Enlightenment into a manifold of contextually dependent rationalities (for example, Toulmin 1989). Confronted with this situation it may look rather strange to apply the concept of rationality to the discussion around technology policy and technology assessment in an emphasised manner.[12] The reason to work with the concept of rationality in spite of these difficulties is that the cernel of the dilemma of technology and environmental policy outlined in part I, seems to be very close to the cernel of the idea of rationality. In the following section 5.1 it will be argued that there might be some "lessons learnt" from philosophy for the central dilemma discussed in this chapter.

The concept of a *pragmatic rationality* is derived from recent work in the tradition of the American Pragmatism (Rescher 1988), the (Continental) methodical constructivism (Gethmann 1996; Hartmann and Janich 1996; 1998) and the work of Habermas (1988a; 1991; 1993). The key idea is, not to postulate an abstract idea of reason and rationality which should be followed by people but to look for rationality criteria inherent to present societies – a culturally-embedded rationality. It is a sceptical approach (compare the discussion in the introductory chapter by John Grin) with respect to the availability of eternally-valid knowledge or fully-founded normative principles, which, simultaneously, insists on the power of argumentation instead of sliding into the arbitrariness of postmodernity. Argumentation, however, implies giving reasons for one's beliefs which are relative to a commonly shared starting point among the participants of the particular discourse. This shall be explained in the following in some detail because it seems to be very important for deeply understanding and solving the central dilemma discussed in this chapter.

---

[12] Up to now, the concept of rationality has been primarily used in these fields in a more pejorative way (Grunwald 1998b, p 29ff.)

5.1
Rationality

A pragmatic reconstruction of the use of the concept of rationality in language (Grunwald 1998b) shows that it represents a category for the assessment of actions and decisions; actions or decisions are assessed for rationality *ex post* or *ex ante*. The purpose of this reflection on rationality is to determine whether "everyone else" in this situation would act (in the ex ante case) or would have acted in exactly the same way as was in fact done (in the ex post case). With regard to the situation of a reflection ex ante this means assessing what the results, impact or side effects of an action or decision would have to be, so that "everyone else" in the given situation would act in the given way or would be able to agree to the decision. The reflection ex ante leads to criteria for decision-making and allows the reasoning for performing a "rational choice": "Rational procedures allow us to arrive at non-arbitrary answers" (Brown 1988, p 35). As such this concept is to be used to designate the *invariance with respect to individual persons* of propositions, actions and decisions. This invariance with respect to individual persons is compatible with pragmatically necessary considerations for contextual dependencies: it does not mean the invariance with respect to *situations*. Pragmatic rationality claims, therefore, do not only *allow contextualisation* but *postulate* them (see below, section 9). Actions are to be designated as rational when they take place on the basis of (descriptive) data and (normative) rationales that can be made to *appear reasonable or justifiable to everyone else* (Rescher 1988; Gethmann 1996). The rationality of actions is based on the trans-subjective validatability of propositions and the justifiability of requests. A *procedural* approach is to be selected for the (cognitive) rationality of propositions, for the (practical) rationality of the suitability of actions to realise aims (the means/end rationality) and the (evaluative) rationality of the aims or purposes themselves (this three-fold distinction has been suggested by Rescher 1988). Rationality is constituted by discursive procedures, for reasoning as well as for justifying: it is constituted by *practical reasoning* (Lorenzen and Schwemmer 1973; Lorenzen 1987; Gethmann 1996; compare also the introductory chapter to this volume by John Grin).

Reflection on the suitability of purpose/action-relationships already provides selection criteria for actions: those who want to achieve a given end cannot simply employ any arbitrary set of means to do so, but rather must select certain means suitable to the purpose in question. Rationality in the sense indicated above, however, extends beyond purpose-related rationality, since it must also take into account the dimension of defining purposes and goals (evaluative rationality) and, as such, also the acceptability of possible side effects, not just the selection of means for specifically defined purposes (cf. Habermas 1987; Rescher 1988). These questions open up the dimension of ethical reflection on the evaluative rationality of purposes and goals as well as on the acceptability of risks (see below, section 7). Ethical reflection belongs to a rationality reflection in technology policy especially if scientific, political and public debates on new technologies find themselves approaching questions such as to what kind of society we *wish* to live in, what self-image of man we are thereby implying[13] and whether, or under what conditions, this, in turn, is desirable. Visions and expectations of the future mostly carry such normative connotations and implications so that a vision assessment, as a rule, should include ethical reflection on its evaluative rationality (Grunwald 1996a, 1999f). This aspect, taking ethical considerations into account and highlighting their relevance in certain cases, distinguishes the concept of rationality used here from a mere technological and scientific ("scientistic") understanding.[14] The rationality of decisions and actions in the sense of their validity "for everyone" requires a reference both to the appropriate selection of means for established purposes as well as to the "reasonable" selection of purposes.

Where are the sources for such a rationality assessment and upon what basis could it be established? The assessment of actions or decisions with respect to their rationality in the form reconstructed above takes place from the perspective of participants in ongoing communication. In the pragmatic concept of rationality the formation of reliable communication and action structures manifests itself amidst the multiplicity of participating players in a society ("Teilnehmerperspektive"; Habermas 1988; Hartmann and Janich 1996). Neither references to systems rationality (Luhmann 1990) nor an exclusive focus on isolated players as is the case in some economic rationality models can be justified on the basis of this pragmatic view. Rationality is always "ours"; whether anyone will consider our decisions to be system rational a hundred years from now may be an interesting historical question when the time comes; but it is totally irrelevant for our present actions and decision-making. This aspect shows that rationality, in this sense, does not refer to some transcendental subject or other instance outside of society. Rationality is a normative concept which *can only be legitimised by society itself*. Any reference to an external agent has become impossible in the post-metaphysical age (Habermas 1988b). The concept of and criteria for rationality are "constructed" by and in society. Therefore, this understanding of rationality is a *culturalistic* and not a transcendental one. It is separated from the transcendental approach of the philosophy of Enlightenment (Kant) by the ideas of the "linguistic turn" of philosophy, by the "pragmatic turn" of the analytic philosophy (Schurz 1988) and the "culturalistic turn" of the constructivistic philosophy (Hartmann and Janich 1996; 1998).

The concept of rationality as proposed here is a normative one *albeit* self-constructed in society. It implies self-obligations and obligations for other people in order to act rationally in the pragmatic sense mentioned above. The normative nature of this concept – the sources of which still have to be analysed, cf. section 7 – makes clear that there is absolutely no substantial presumption that humans *are*

---

[13] In the paper of Rob Reuzel and Gert van der Wilt there is an example how such a concept of Man was – in the reconstructive perspective - implicitly presupposed (to be free from handicaps if ever possible) which leads to a failure of the technology (CI) because this concept is not shared by the persons affected.

[14] It has been the main reproach of the Critical Theory against the concept of rationality that it were reduced to an instrumental understanding of reason. In Habermas' words the „technically halved rationality" was criticized as a reductionistic one. In this contribution, instead, an emphatic concept of rationality without such reductionisms is favoured.

in fact rational beings or, in fact, always act rationally.[15] The „rationality of emotions", often mentioned to criticise pure rationalism, is neither ignored nor ruled out in this approach. Of course, humans often are emotional (with emotions sometimes leading to conflicts with rationality requirements in the sense mentioned above). But the point is that emotions *per se* do not replace arguments, they are neither rational nor irrational – there may be very good arguments or very bad ones behind them. To check emotional expressions with respect to their argumentational tenability rationality assessment procedures like discourses are indispensable (Habermas 1991; Gethmann 1996). The task of such procedures is to transcend the sphere of mere individual beliefs or emotions with respect to their *generalisability* – an eminent important factor to assess the validity of emotions for "everyone" and, as a consequence, for the opportunity to base legitimate decisions on such emotions.

In talking about technological development in the 20th or for the 21st century there have been and still are many fears or feelings of being uncomfortable. These fears or feelings have to be taken into account very carefully; but they cannot be the only foundation for decisions on how to shape future technology because they are *merely subjective*. A rationality analysis should – without any embarrassment – uncover if there are more or less tenable *trans-subjective* arguments supporting these emotions. In this way, rationality is a *self-construct of society for the purpose of transcending individual beliefs towards trans-subjectivity in order to arrive at well-founded collective actions and decisions* (Grunwald 1999a). It is used to optimise the possibility to anticipate the success of collectively relevant actions and decisions, to allow a balance between continuity and flexibility in society (cf. section 5.2), to use predictions as planning data and to enable societal learning processes (cf. section 5.4).

Because the concept of rationality itself is a normative one the criteria of rationality are to be justified very carefully in order to ensure the validity of conclusions based upon them. The basis for justifications can – due to the culturalistic approach – only be found in society, by uncovering and reconstructing its underlying normative structure (Gethmann 1992; Habermas 1991; Grunwald 1999f). The assessment of actions, plans, technological options or visions with respect to rationality standards must be based on *a pre-discursive agreement factually accepted by the people or groups concerned* (Gethmann 1982, Grunwald 1999f). This type of acceptance-orientation is indispensable because without acceptance of such very elementary issues there couldn't be any rational and legitimate basis for assessments with respect to social continuity and long-term issues (section 7). Pre-discursive agreements relate to the life-world (Gethmann 1992) and are relatively stable pre-conditions of discourses. Rationality criteria in this way are acknowledged *implicitly* in the practices of society – which does not imply that they are always accepted *explicitly*. This field of tension between implicit and explicit acceptance is very important in dealing with the key question the present chapter is devoted to (section 7).

---

[15] *Substantial* rationality postulates are inherent, for example, to the homo oeconomicus model of some economic rationality theories.

> **Box 6:** *As an example, consider the problem of taking responsibility for future generations in the field of global warming. This type of responsibility is, in principle, implicitly agreed upon in society, by the customs of herediting one's property to the children or grandchildren or by the duty of the state to care about the affairs of future people ("Daseinsvorsorge"). The factual acceptance of this obligation may then be used to motivate long-term responsibilities in other fields than the traditionally established ones. The task is not to motivate a completely new type of ethics (Jonas 1979) but to look for and to uncover the relevant, implicitly acknowledged morals and to transfer them to the newly oc-curing challenges.*

This structure of rationality may be imagined as follows (thanks to Rob Reuzel for this clarifying picture): individuals should first remember their commitment to rationality in the sense of being obliged by something like a *social contract* defining the procedures of constituting the "common good". These self-obligations imply looking beyond short-sighted self-interests and serve as a pre-discursive agreement to perform discourses around the actual topics of interest (for example for decisions about certain long-term agendas), focusing on acceptability criteria that may guide collective action rather than on acceptance of a given vision.

One may wonder why people should agree at all to this social contract committing them to look beyond mere self-interests. This question may lead to the impression that people could, indeed, choose between the options 'contract' or 'no contract'. In my belief, however, there is *no real option like this*. We are all living in traditions and bound to the past in the sense that there is no arbitrariness in choosing things like the underlying social contract (Grunwald 1999f). We are able to modify the contract incrementally (Kettner 1999 and the literature mentioned there) but we cannot switch *ad hoc* to a completely different one (except, perhaps, in very rare revolutionary situations). The question, therefore, is not *why* people should acknowledge such a contract but *how to make the contract and its implications explicit* and *how to deal with its consequences* by the people affected.

Rationality reflection ex ante is an action- and decision-related concept, one of its purposes being the creation of (hypothetic) collective agreement for decision-making purposes. The relation of the concept of rationality and the problems of technology policy mentioned above should become clear at this point. Rationality reflection enables collective actions or actions with collective impact "non-coercively" in a pluralistic society because the descriptive and prescriptive premises of the decision are validated and justified and, therefore, should be understandable and agreeable for "everyone". This *pragmatic* understanding of rationality leads to consequences relevant for many questions of policy and decision-making. Two points which are closely related to the discussion in the contributions to this book around expectations, visions and the role of technology assessment, are to be highlighted, (1) the difference between rationality and success and (2) the relation to philosophical universalism.

(1) By means of a rationality assessment of a decision under consideration the *probability of success* of that decision can be optimised. Rationality assessments ex ante serve the purpose of creating the *best possible reasoned* expec-

tation of success relative to existing knowledge. The success of decisions and actions, however, cannot be *guaranteed* through rationality reflection ex ante (Rescher 1988; Grunwald 1999a). Rationality and success are quite different assessment concepts for actions and decisions. While the success of an action can only be proven after the action has been undertaken, it is precisely the purpose of rationality assessments ex ante to determine rationality and to distinguish rational and less rational or irrational decisions from one another *before* success and failure can be determined. It is the case that a plan which has been proven to be rational may not necessarily be successful and that an "irrational" plan may be successful in spite of its irrationality (Grunwald 1999a).

Rationality, therefore, must not be misused to suggest a *guarantee to success*: also acting rationally is acting under risky conditions. Rationality considerations cannot erase the risks arising from uncomplete knowledge, changing goals and unpredictable contextual conditions. Consequently, the Cartesian rationality standards characteristic for High Modernity (Toulmin 1989; compare the introduction to this volume by John Grin) are not justifiable. They overstress the security and guarantee aspects of actions and decisions (which had a revival in the planning discussion in the fifties and sixties of our century, Grunwald 1999a) and they neglect the risks indispensably involved. In this way, considering actions without reflecting their risks is not a very rational action scheme. Rationality assessments shall just point to the risks involved and to reflections on ways of dealing with them. The reproach to the Cartesian paradigm in the sense mentioned by Toulmin should not be that there was too much rationality in it but, on the contrary, that it *shows rationality deficits ignoring the pragmatic side of rationality*. It includes too little rationality because the strong security postulates themselves are *not justifiable but dogmatic* ones.

(2) There is no strong philosophical universalism involved in this pragmatic model of rationality though it may sometimes look like this. The "validity for everyone" included as an essential in this model leaves open who shall be understood by "everyone". In the universalistic interpretation "everyone" stands for the whole of humankind including past and future generations (Habermas, Apel). In the pragmatic concept of rationality the term "everyone" itself must be defined pragmatically and contextually. From an ethical point of view (Grunwald 1999f) any person concerned has to be taken into account: everyone affected by a certain action and its consequences has, in principle, to agree upon this action. The set of concerned people, thus, may reach from a small group of persons (e.g. for actions within a family's area) up to all present and future generations. This set, obviously, *depends strongly on contextual conditions and the problemtypes involved.*

> **Box 7:** *In the problem of global warming, obviously the presumed interests of future generations have to been taken into account. The question then is whether all future generations should be handled equally of if it is allowed to shift some burdenings to the future by introducing discount rates in calculating the measures required at present.*

These points both illustrate the main differences between the pragmatic and the (untenable) „certistic" rationality concept governing Cartesian philosophy: also rational actions and decisions are in risk of failure and the claim for rationality includes strong requirements for contextualisation. Vice versa, the pragmatic rationality concept maintains the request for normativity by pointing to the validity of rationality standards for "everyone" and to the *non-arbitrariness* of actions and decisions. In this way, the pragmatic claim for rationality is weaker than within the Cartesain paradigm but stronger than within a purely, a-normative, skepticist paradigm. It allows and demands contextualisation (see below) *without* abandoning the idea of rationality standards valid for everyone in a pragmatic sense.

5.2
Rationality and the Balance between Continuity and Flexibility

Choosing the *invariance with respect to individual persons* (Rescher 1988, Gethmann 1996) as the crux criterion of rationality enables, as a consequence, some conclusions to be derived for the dimension of collective acting in time. Within a culture it is impossible to act in a complete arbitrary way if the action shall be designated to be rational. The traditions, rules, morals, laws and other customs constitute a normative framework for actions and decisions (like a "corridor" of rationality). An action of a certain person is accepted as "rational", if it is "understood" by others using criteria from this cultural framework. Exactly in this case the invariance postulate of the "understandability for everyone" is fulfilled ("everyone" in the pragmatic sense mentioned above). If the actions of a certain person are "understood" by someone else according to criteria derived from the framework an inductive conclusion on the understandability for everyone within that cultural framework is allowed. If a culture is expected to be internally and externally recognisable as a culture – this expectation seems to be self-evident, derived from the concept of culture (Gutmann 1998) – there is a need for a certain amount of stability and continuity of the normative framework allowing such rationality assessments. The understandability of actions and decisions of other people is lost in the case of "irrational" actions. Simultaneously, as an analytic conclusion, the pragmatic concept of rationality is a contextualised one: what actions and decisions are deemed rational depends on the cultural embedding and the surrounding context of that rationality assessment.[16]

---

[16] This aspect of rationality is the main reason for the demand for contextualisation in technology assessment (John Grin in this volume). In the same way, it shows the deficits of the Cartesian concept of rationality neglecting these requirements for taking into account the contexts in any

The phenomenon of "understandability" is the source of rationality assessments; it ensures the *condition of the possibility of a culture*. If we were acting in a completely arbitrary way, there could not be any culture but only solipsistic individuals who could not understand each another. In this sense, rationality is a key concept for understanding the kernel of cultures.[17] Rationality in this way is constructed by society by reflecting its conditions for the common understandability of actions and decisions and making them explicit. Regarding decisions with great impact on future, their rationality ensures that these decisions can be "understood" in future, too (not, of course, in an arbitrary far future, when the cultural conditions may have changed dramatically, but in a more or less near future).[18] This is exactly the point where the long-term planning requirements (section 2) and the concept of rationality meet each other: rationality constitutes the medium of society to enable continuity, stability and identity beyond the individual level.

Accordingly, the concept of rationality as a self-construct of society is suitable to deal with aspects of shaping the future. This can be viewed, at least, in two directions. At first, introducing the concept of cultural rationality as above implies that a certain kind of *continuity* must be guaranteed if rationality is claimed. Secondly, in spite of such indispensable issues of cultural continuity all affairs in culture are evolving in fact: the purposes and values followed in society, the knowledge available, the situations and contexts of decision-making etc. Any decision-making which claims to be based on rationality has to take into account such modifications and innovations (this is one aspect of the contextualisation required as mentioned above) and, thus, leads to strong requirements for adaptation and flexibility (Grunwald 1998b, section 6 below).

Society cannot decide either for continuity or for flexibility: in shaping technology, both flexibility and continuity are required and have to be included in rationality assessments. For practical reasons it may often be unavoidable to close down or to "freeze" action plans or steering policies or certain parts of them. Optimal goal attainment, however, requires that latitude be kept open for flexible adjustment to new situations between the time when the decision is taken and the time when it is implemented or during implementation. "One could call this procedure rationality with explicit rules in cases of error" (Bechmann 1991, p 70; translation A.G.). The gain of rationality consists in the fact that any change must and can be reflected with respect to its relations to former versions and to the newly occurring situation requirements; this rules out merely trotting along behind contingent and short-lived fad phenomena. As such, the concept of *flexible planning* (Grunwald 1999a) is of outstanding importance for the formulation of reliable long-term tenets of technology and science policy as well as for the necessary

---

assessment very carefully. Consequently, it comes out that this concept of High Modernity *lacks* rationality because of dogmatic and non-justifiable presumptions.

[17] The normativity preventing us from acting in a completely arbitrary way is exactly what I mean by the implicit social contract in which we are embedded and to which we are bound.

[18] Historicians are often confronted with the principal problem that they have at first to uncover the rationality standards of the historical context under consideration before they are able to interpret the historical documents (the well-known hermeneutic circle). For the methodological problems involved in reconstructing past planning processes compare Grunwald 1999a.

balancing of short-term flexibility requirements and long-term stability of a society, taking into account its cultural identity and its traditions as the basis of rationality assessments.

Special importance is attached to the creation of planning security for the players involved (technology manufacturers and consumers). As such, the *ex ante* distinction of rational and less rational or irrational options in decision-making situations serves as a stabilising way for society to deal with issues relating to the future between continuity and flexibility. Options for the future can be analysed in terms of their rationality, *i.e.* the extent to which they should be acceptable for everyone (in particular for future generations). This will make them viable over the long term, above and beyond considerations based on merely situational opportunity.

In conclusion to this subsection it shall be emphasised that society has to balance carefully between the (often diverging) requirements for long-term continuity and short-term flexibility. This balancing seems to be the real challenge for society, instead of purely accelerating all social processes according to the increasing speed of technological change.[19] Such a balance – which has to deal with aspects of rationality and stability of society – cannot be based on pure acceptance (section 4).

5.3
Rationality in Forecasting and Prediction

Technology policy is always concerned with future aspects, either in issues of shaping future technology or in forecasts allowing early reaction to innovative developments. Therefore, assessments in technology policy need to answer questions as to in what way prognostic statements are to be understood, in what way prognostic statements are to be approached and the dilemmas inherent to predicting technology development and its impact (Peter Mambrey in this volume; Banse and Friedrich 1996) shall be dealt with.

The rationality assessment of predictions questions their argumentative validity "for everyone". The fact of the matter is that there is no chance to succeed in accurately predicting what the real situation will be in the future; it turns out that the rationality of predictions does not mean that they will actually turn out to be true (Bechmann 1994; Renn 1996; Grunwald and Langenbach 1998). As such, the objective in technology assessment cannot be that of making "correct forecasts", given that it cannot be determined *ex ante* whether or not a prognosis is correct - one has to wait whether it becomes true or not. This is simply an application of distinguishing between rationality and success (section 5.1): rationally founded predictions may fail, irrational ones may correctly foresee future events. This

---

[19] This is also valid for the present discussion around shaping the information society. The decreasing rate of knowledge growing old does not imply that cultural and political affairs have to be simply adapted to this development. On the contrary, policy regulations are required to ensure the relative stability of cultural contexts facing the rapid variability of knowledge and technology in this field.

means that it cannot be the main sense of predictions to prognosticate future events or processes as they will really come out to be true.

The task of prognoses is, instead, to rationally analyse the hypothetical consequences of decisions to deliver data which can be used for decision-making and planning (Grunwald and Langenbach 1998). Even well-founded predictive assertions remain hypothetical because of their foundation structure. Any prediction relies on a *relevance decision* made *before* predicting by which the boundaries of the "prediction system" are chosen. Within this system the parameters and their interdependencies are analysed to allow predicting. The "world" outside is *excluded* from this consideration. This prevailing distinction between relevant and irrelevant aspects of a given problem obviously is, in principle, fallible: there may be relevant factors being left outside as is often a main reason for the failure of predictions.[20] Therefore, a guarantee for "successful" predictions cannot be given already for purely methodological reasons. Substantial assumptions about the assumed complexity of the world, non-linear dynamics of society or similar statements empirically provable – as noted by Renn (1996) – are not necessary in supporting this conclusion.

The rationality assessment of predictions, prognoses and forecasts, therefore, is concerned with the argumentation structure and the risks, limitations and restrictions hidden in the argumentative chain leading to them. Especially the normative presuppositions are important: decisions regarding relevance, which are required *before* the prediction can be made, are often dependent on normative predecisions. Therefore, forecasts and foresights are often, though they seem to be purely descriptive, *carrying normative implications*. Indeed, even (seemingly descriptive) predictions are not free from elements of (normative) *shaping*.[21]

This methodological reflection makes it possible to shed light on some aspects of the *prediction dilemma* in technology assessment (Bechmann 1994; Banse and Friedrich 1996; Renn 1996; Grunwald and Langenbach 1998). Eventually, if the future were foreseeable it could not really be shaped. Indeed, in some (mainly earlier) concepts of technology assessment a certain kind of "technological determinism" was assumed. It should allow predictions of "real" developments in the future following the (deterministic) laws of technology development. In this way, the resulting predictions should help society, especially its political system, to prepare for the ongoing and presumably pre-determined developments in order to adapt the fields concerned in society to these developments. Technology was seen as the main driving force in society forcing the political system to set up adaptation processes early enough (the "early warning" function of TA, cf. Bechmann 1994). The underlying determinism, however, is not well-founded as has been

---

[20] This often applies to trend extrapolations assuming (due to an explicit or implicit relevance decision) some *ceteris paribus* conditions to be fulfilled, for instance in the way that the established trend might simply be extended to the future. Consider, as an example, the predictions of the future energy consumption made in the early seventies simply extending the increasing consumption rates of the past to the future. The underlying assumptions have been proven false by energy saving strategies decoupling economic growth from energy consumption.

[21] This becomes very obvious in the well-known cases of self-destroying or self-fulfilling prophecies, cf. Grunwald 1999a.

shown by many sociological field studies (cf. the contributions in Bijker et al. 1987; Dierkes et al. 1992; Bijker and Law 1994; Rip et al. 1995a) as well as by theoretical work (for example, Grunwald 1998c and the literature mentioned there). The prediction dilemma, therefore, isn't really a dilemma. It just expresses exactly the fact that there cannot be any determinism allowing sure prediction, which is compatible with pragmatic rationality.

The problem, therefore, is not that we know so little about the future, as is often lamented. We should, indeed, completely erase the idea of a Delphi oracle because of its underlying presuppositions of a deterministic world. The "risk society" (Beck 1992) is, in turn, a type of society shaping its future by using its own mechanisms and procedures without any external reference. The real problem, instead, is how to come to trans-subjectively valid ideas on *how the future should be.*[22] Rationally shaping the future is possible on the basis of a collective focus on goals established as well as on available trans-subjective knowledge on which to base actions and prognoses. *Predictions do not anticipate the future, but they constitute planning and decision data* (Grunwald and Langenbach 1998, p 107ff.).

As a side-effect, it comes out that there *cannot be any experts for the future.* The future is only accessible by just performing actions and by decision-making. There is no expertise available on what the future will really look like. The so-called future experts like trend analysts and observers of the advance of science and technology are experts for certain aspects of the present situation and of the present expectations for the future, not of the future itself. The question is not if the future is shapable at all because there is no other access to the future than shaping (within boundary conditions and the cultural framework). The alternative options are not "to shape" or "not to shape"; there are only different ways *how to shape.*

Box 8: *Consider the global warming example. Independent of what political or technological measures are implemented at the present time (including the option of not reacting to recent climate predictions at all) the future always will be shaped by present decisions. These decisions open up certain "futures" and close down others.*

5.4
Rationality and Learning

The possibility of learning and rationality assessments are interconnected in that only statements for which arguments can be given can be improved. If it were not

---

[22] This postulate should not be misunderstood in the way as if it would be presupposed that future could be shaped *in an arbitrary way*, free from any constraints. Indeed, there are many boundary conditions not open for real decisions and shaping. As an example, we are not free to decide if we wish society to transform into an information society or not; the opportunity for shaping in this field, however, appears at a different level: in what way aspects of the information society are embedded into the cultural framework of society (Grunwald 1999g).

possible to base actions on rational criteria, the only means of learning would be to use the (extremely inefficient) method of trial and error. Rational reflection *ex ante* thus the basis for reflection *ex post* on the reasons for success or failure and forms a necessary condition for learning (Habermas 1988a, Grunwald 1999a).

If the action leads to success the assumptions involved are proven to be suitable. In the opposite case the site in the argumentation structure might be identified where a false or unsuitable assumption caused the failure. Knowledge improved in such a way could be used in the next decision-making situation involving this type of knowledge to improve the expectability of success by avoiding this type of failure. *Vice versa*, if one would, as an example, rely only on unfounded prophecies there could not be any chance for learning, neither in the case of success (which, indeed, might occur though its unfoundedness, see above) nor in the case of failure. There could not be any "lessons learnt" for the next case. In this way, such cognitive issues are important for learning. More generally speaking, learning consists of a special type of communication between various actors (Grin and van der Graaf 1996) with regard to certain cognitive mechanisms.[23]

6   Incremental Technology Policy Without Losing Orientation

The next challenge consists in relating the general rationality considerations, especially concerning the analysis of the role of continuity and flexibility in society (section 5.2), to the central dilemma of this chapter. How to combine the issues of flexibility (required, for example, by acceptance aspects) and the long-term stability (required by many decision situations, section 2)? An approach to technology policy is proposed in this section which is flexible and incremental in performing the detailed steps of decision-making but avoids the shortcomings of a *purely* incrementalistic approach (section 4). At the same time, it allows visions and long-term considerations to be sustained over a relevant amount of time – if certain conditions are fulfilled (section 9).

6.1
Technology Development and the Role of Technology Policy

In the following a closer look is to be taken at the possibilities of reflection on rationality in the process of shaping technology in society. Actions and decisions of individual engineers and scientists as the producers and the users of technology will not be discussed here. Neither will institution-level decisions concerning technology-relevant companies (for products and product lines or for development process and production processes) be dealt with. Instead, the focus here is *on societal or governmental mechanisms* for steering technology (political actions at the

---

[23] The role of cognitive issues should not be under-emphasised in actor-focussed approaches. Learning is a social behaviour as well as communication with regard to a *cognitive* disposition.

government level). More specifically, these mechanisms can be classified further in order of decreasing depth of government intervention:

a) direct government-operated or at least government-dominated technology development (aerospace, military technology, nuclear power plant technology, transport infrastructures, etc.),
b) indirect government steering of technology through research promotion and technology-support programmes,
c) indirect government steering of technology through the regulation of general conditions for technology development in the economy.

Government steering of technology through direct regulation (legislation or other forms of regulation) creates framework conditions for general technology development which, under these conditions, takes place in the economy under conditions of competition. Societal or political activities of this kind relevant for the shaping of technology are, for instance, the establishment of emission standards, other environmental standards, safety standards, the establishment of technology-relevant tax rates, technology-relevant and direct regulatory measures such as take-back requirements for old cars or other recycling requirements. Regulation of the process of shaping technology in this sense either imposes strict limits (e.g. ban on new registration of cars without catalytic converters, ban on human cloning) or formulates the framework conditions for market behaviour such that development in the desired direction takes place for economic reasons (e.g. pollution-related gradation of the motor vehicle tax, or a possible energy tax). The process of shaping technology through direct regulation is a reflected process of shaping framework conditions for further technology development.

Governmental and societal measures to control technology development do not fully define the degrees of freedom in the process of shaping technology. They constitute boundary markers. Shaping technology within the established limits of the law requires that these are respected; however the respect of these limits in itself does not provide sufficient guidance. They merely provide a negative definition of the direction in which development is supposed to go by acting as a selective filter for technologies which it is felt are undesirable. What is involved is a "technology-open" regulation relating the boundaries of technology development to societally followed and accepted visions and goals for the future.

Beyond this level of regulation in which rational analysis always extends to conditions and consequences of political action, the process of shaping technology in actual fact takes place under market conditions. After all engineers and scientists bear special responsibilities as individuals in connection with their direct involvement in the processes of developing, producing, using, and safely disposing of technologies. The "rationalities" of these players are, however, particular and not general; the understandability of their actions for "everyone" cannot be claimed (Grunwald 1999d, 1999e, 1999f). The rationality they are obliged to due to their professionality is a mere economic or technical one. The level of emphatic reflection on rationality in the sense mentioned above is, thus, only twice: that of government steering of technology development ("commitment to the public good") and the level of citizens (citoyen) constituting the public.

This does not only apply for reasons of division of labour in the process of shaping technology development but there are also legitimacy aspects to be considered. The imposition of limits or the setting of goals for technology by society only then makes sense if and when this can be made generally binding. Neither companies, engineers, nor technology users have a legitimate right to declare their individual or sectoral views as goals for society in general or as general limits for technological development. This can be done only in the context of legitimate policy and decision-making procedures. That is why this is a place for comprehensive reflection on rationality in accordance with the criterion of understandability "for everyone". Once this is established the next step is that of defining the type of rationality to be applied in the process of shaping technology development.

6.2
Planning Rationality in Technology Policy

All the three levels of shaping technology in society mentioned above can be viewed from the standpoint of planning rationality criteria, but as noted, we will focus on the government level from the standpoint of shaping government framework conditions for technology development. This third level is to be focused on in the following.

Technology policy decisions serve the purpose of prestructuring the future by promoting or selecting individual options or entire bundles of options out of the existing diversity of possibilities that exists and in doing so steer technology development in certain directions or rule out certain other directions. By seeking to exert specific influence on the potential of future players to undertake actions, what is involved in connection with these policy activities is *planning in the general sense of an intended exertion of influence on future potentials for action* (Grunwald 1998b; 1999a; 1999c). What is involved here is not the planning of technological products or technology development as a whole in the sense of determining further development – i.e. not a "planned economy" ("Planwirtschaft") – but rather the rationally thought out shaping of framework conditions for further technological development.

Reflection on these framework conditions from the rationality standpoint leads to the concept of *rationality in planning*. As such, rationality in shaping technology can be understood as rationality in planning. In our complex societies decisions are made the purpose of which is to achieve optimum effects with limited resources under predefined and justified criteria. Rationality in planning and decision-making is thus by no means outdated; but its use in dealing with problems of technology development need to take into account the realities of modern society. That is:

(1) Planning in this connection does not mean the detailed planning of actions, nor does it mean an algorithm-like process of carrying out predefined steps; what it means is the planning of objectives with varying depths of planning (Grunwald 1998b, 1999a). The depth of planning is itself open to the process of critical and contextualised reflection.

(2) While, has just been noted, there is reason to reject planning scepticism, there is also a need to reject the philospohical position of planning euphoria (Grunwald 1999a). Viewing the future from the standpoint of planning rationality must not lead to an assumption of the ability to create pre-figured and fixed future conditions. Planning, too, constitutes an activity *involving risk.*

(3) This means in particular that rational steering of technology development, such as in the field of research funding or government regulation, must always be temporary planning since by definition it is *flexible planning*. Specifically, it must be temporary in terms of normative premises, the knowledge used, the objectives pursued and the contexts to be taken into account (sect. 5.2).

Shaping technology relative to criteria of planning rationality cannot lead to predetermined final states but can only consist of acting under the permanent obligation to reflect (the normative premises, the state of knowledge, the purposes and the interpretation of the relevant context aspects). Shaping the boundary conditions for technology development being the task of technology policy, the task of TA should consist of this permanent reflection under aspects of practical, evaluative and cognitive rationality (Rescher 1988). Shortening rationality to descriptive aspects is impossible if the shaping of the future is considered under aspects of planning rationality. The quality and reason of plans depends on the technical rationality in terms of means and ends as well as on the ethically justified choice of reasonable purposes, expectations and visions.

6.3
In What Sense is Technology in Society Malleable?

Flexibility requirements in R&D processes can be focussed on certain points of an anticipated line of development by defining so-called milestones. Such milestones are branches in the decision-making tree at which the fulfilment of certain intermediary steps is to be checked in order to decide about alternative ways of proceeding or of modifying the initial plan. This mechanism might be adopted by technology policy. *Flexible rational planning* (Grunwald 1998b; 1999a) allows *reflected adherence to lines of action in a changing environment* with a view to achieving objectives in the face of limited societal resources. This shall be explained in the following by contrasting this approach with two other ones:

- the technocratic planning model dominant in the sixties (cf. the discussion in Camhis 1979),
- the incrementalistic model proposed by Popper and Albert (cf. the discussion in Camhis 1979 and Grunwald 1999a) which's dynamics seem to meet very well the dynamics of the acceptance-oriented approach criticised in section 4.

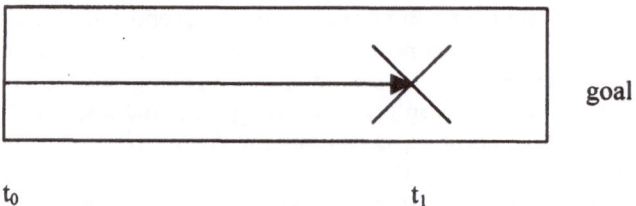

goal

t₀   t₁

Fig. 1. technocratic planning procedure, departing at the time t_0 and terminating at the time t_1 when a pre-determined envisaged goal of executing the action chain straightforwardly is reached.

The technocratic planning approach (fig. 1) presumes the plannability of large areas of society at the macro-level with – in the ideal case – a guarantee for success. That means if such a plan – based on sociotechnological research on certain laws guiding the development of society – would be brought to practice at a certain time t_0 it is assumed that its goal will be reached at the time t_1 scheduled before. Such a technocratic understanding of the dynamics of social evolution has proven, in the meantime, to be unfeasible because of the non-availability of deterministic social laws. Also, it does not meet the dynamics of society. The second main point of criticism is that this model is designed to the state as a central planning agent. This role, however, cannot be fulfilled by the state anymore (van Gunsteren 1976; Gottschalk and Elstner 1997). Thirdly, such a model is simply not desirable because of the occurring problems with any determinism (compare the discussion that there are no experts about the future in section 5.3). Society, therefore is not plannable at the macro-level. Furthermore, to talk about the plannability of society as a whole leads to the wrong questions (Grunwald 1999a).

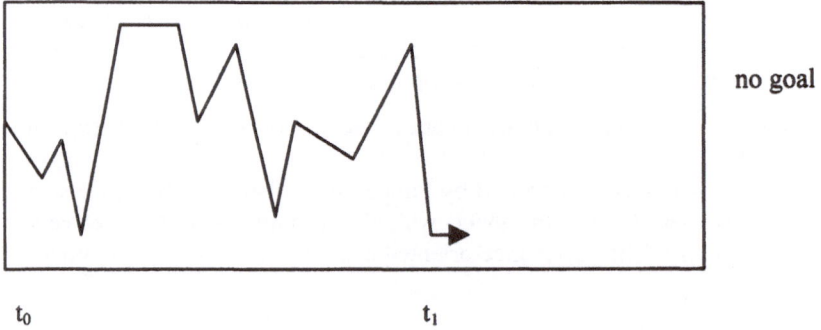

no goal

t₀   t₁

Fig. 2. Purely incrementalistic planning procedure, departing at the time t_0 and terminating at the time t_1.

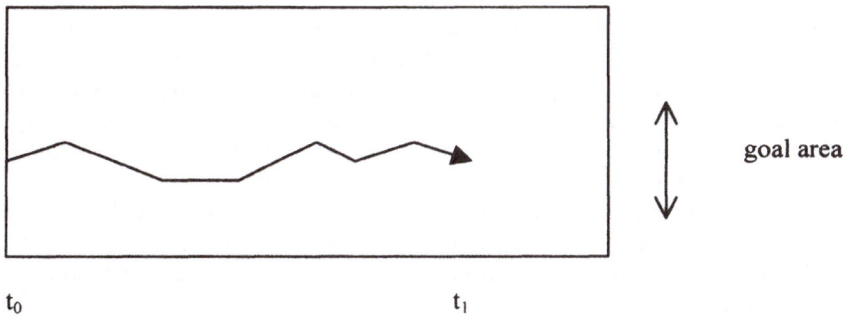

t_0 t_1

Fig. 3. *Directed* incremental planning procedure, departing at the time t_0 and terminating at the time t_1.

The incrementalistic approach (fig. 2) due to the possibly rapid-varying acceptance level has extensively been discussed in section 4 (disjointed incrementalism, Braybrooke and Lindblom 1963; Camhis 1979). The dynamics of such an approach are resulting in possibly erratic changes and in a complete loss of the direction of social development. What state is reached at the time t_1 is widely dependent on contingent events and on processes dominating the rate of acceptance of that development at the several „presences" the process has to pass until reaching t_1. This approach skips the challenge of shaping society and technology, it dissolves any attempt for shaping into a mere evolutionary process without any normativity (Camhis 1979; Grunwald 1999a).

Within "directed incrementalism" resulting from the considerations and reflections above, the direction of action and decision is maintained but not fixed. Permanent reflection on the goals and the means to attain them lead to incremental changes of direction in the development, of the goals as well as of the measures to reach the goals. This change, however, does not occur upon the basis of chance events and does not show an erratic behaviour. This kind of development allows to get closer to the envisaged area of goals and to take into account the short-ranged flexibility requirements (which are leading to the incremental changes of direction).

Disjointed incrementalism and technocratic planning can be expressed as being the extreme positions on a continuum of possible planning strategies. Directed incrementalism, falling some place in between, fits very well to the concept of pragmatic rationality and seems to be very suitable to explain technology development as well as to serve as a model for shaping technology by many small but reflected steps (in the same sense, compare also Rip et al. 1995b, p 8; Schwarz and Thompson 1990). The rationality reflection in this context allows long-term directions to be maintained though the method of proceeding is the incremental one allowing short-ranged flexibility requirements to be taken into account (for example, caused by acceptance problems). In this way, shaping the future is imagined by making many small and reflected decisions under guiding visions.

> **Box 9:** *For example, consider the former plan of the Green Party in Germany to increase incrementally the fuel taxes to enforce a reduction of the consumption of fossil fuel.[24] Such a model gives long-term orientation and ensures planning security for car owners and users as well as for the automobile industry. In this way it allows a long-term agenda to be followed. On the other hand, the process of incrementally and stepwise increasing the fuel tax allows learning processes: modifications in order to slow down or to speed up the increasing process due to experiences gained during the implementation phase are possible. Though the direction of the process is clear the implementation allows learning and flexibility.*

One problem still remains: how to normatively ground the direction of development? This question shall be dealt with by dealing, in the last part, with the concept of acceptability.

Part III   Resolving the Dilemma - as far as Possible

7   From Acceptance to Acceptability

What should be done – in order to return to the dilemma outlined in part I - to take into account the people's view without sliding into the shortcomings of a mere acceptance-orientation? Summarising briefly, we have to acknowledge that *without* respect to acceptance at all, sustainable development or stable technology policies cannot be reached in a democratic and pluralistic society. Relying *exclusively or mainly* on acceptance, however, does not allow long-term plans to be followed. The way suggested to deal with this situation is to *shift the level of acceptance* required. The direction of solving the dilemma consists of acceptability criteria and procedures to create them. The challenge is, at this point of the analysis, to correlate this approach with the questions left from Part II and to answer the question how to ground acceptability criteria and the procedures leading to them.

What someone accepts is first of all his or her subjective decision according to the individual preferences and is, thus, pretheoretical.[25] Technological actions and decisions, however, may have consequences for "everyone", which is why "everyone" should have a chance to agree to them (for ethical reasons and because otherwise conflicts could be generated with the risk of blocking decision-making processes in society). The question, therefore, as to what extent it is conscionable to *expect someone to accept a technological development* is of wider societal and political interest. Rationality assessments are not based on factual acceptance at the level of individuals but on the *acceptability* of decisions at the societal level. Rational technology policy, therefore, should be based on justified criteria of ac-

---

[24] This plan shall only be used to explaing the essentials of the directed incrementalism. It is not intended to support or reject this plan.

[25] Factual acceptance in itself is neither rational nor irrational but a-rational. Its rationality can only be checked by procedural analysis and rationality assessment (section 5).

ceptability (for the field of establishing environmental standards compare Renn and Pinkau 1998).

The basic idea is that there are *implicitly* accepted norms and presuppositions in society (elements of the „social contract") which can be used to formulate criteria for acceptability. The method to extract these criteria should be to uncover and reconstruct the rationality in current practices and culture (Habermas 1991; Gethmann 1992). The question then is no longer what technology will be accepted but what technology and technology policy *should be acceptable* according to such underlying normative presuppositions of society and to its inherent rationality standards. In order to justify rationality criteria for *acceptability,* reference to a *preceding consensus* shared by the decision-makers and stakeholders is required as a basis for argumentation and negotiations. Such a consensus must be agreed upon. This acceptance-orientation is indispensable because otherwise there wouldn't be any rational basis for argumentation and reflection on planning and decision-making, for social continuity and long-term issues.

Since preceding agreements relate to the life world and are thus relatively stable pre-conditions of societal decision-making mechanisms, reference to their acceptance, in contrast to the direct acceptance of technologies, does not lead to a dependence on short-term and chance events. The level of acceptance required is shifted from the acceptance of the factual technology and technology policy *to the acceptance of the rationally justified criteria and procedures for decision-making.* If the level of "having to be accepted" is shifted to lifeworld presuppositions, the benefit of using the concept of cultural rationality consists in the stabilisation of the basis for assessments and decision-making; in contrast to short-term trends, cultural presuppositions are relatively stable over the long term (Grunwald 1998b).[26]

Rationality Assessments of technological options, therefore, do not consist of an empirical investigation of the presumed acceptance but of establishing a normative *acceptability* of decisions in the above sense. For society it is particularly of interest to ask to what extent certain persons and groups *can be expected* to accept a technological development in the interests of society as a whole. Furthermore, it often has to be decided how far the "common good" legitimately allows people to be burdened with consequences resulting from reasonable and legitimate long-term considerations.

Take as an example the preceding consensus required by ethical reflection (Grunwald 1999f). Why, actually, should moral conflicts be resolved at all by ethical reflection on the basis of *arguments* and not by other means? Briefly, the analysis of this question leads to the concept of human rights, human dignity and the moral autonomy of persons. However, in spite of this emphasis on Kantian elements of ethics the idea of the factual autonomy and freedom of the individual, i.e. the transcendental subject as it were, need *not* (!) be accepted. The preceding consensus referred to does not imply any assumption of a *factual autonomy* of the

---

[26] Obviously, such cultural foundations are varying, too. For example, the necessity for environmental protection has belonged to the Western culture for some decades, before this time the "social consciousness" was not aware of this topic. These processes, however, develop smoothly and slowly, they do not lead to *short-ranged* dramatic changes.

individual. This autonomy is "only" *presupposed in a counterfactual sense*: Ethics is practised in a modus in which the players in question *are assumed* to be acting as free individuals – and if this were not the case then the provision of ethical advice would be obsolete in practice. Ethics is possible and relevant in practice if these counterfactual assumptions are not mere figments of the philosophical imagination, but rather established in practice and actually effective there. The factual in the counterfactual – i.e. the *factual acceptance of the counterfactual elements of ethical justification chains* – must be found in concrete societal practice (Habermas 1991). It must be more than just a mere construct, but has to be genuinely represented – e.g. by voters wanting to be taken seriously as persons who play an active role and not just as statistical material, or by assuming in the legal system a defendant's "ability to be guilty". Counterfactual elements of ethics as assumptions with regard to the moral recognition of persons and argument-based communication about conflicts are, in the final analysis, utopian ideas about "what life should be like", *i.e.* about the ultimate objectives of political action. These counterfactual assumptions must be factual even as counterfactuals, i.e. they must be part of the 'social contract'.[27] As such, ethics has an element of *acceptance*, otherwise it would lead to normativistic fallacies.

Transferred to the general reflection on rationality this means that rationality criteria derived from the understandability "for everyone" must have their basis in the factual elements of society, if technology assessments based on them are to be able to lay claim to acceptance and implementation. Proving this in detail will continue to be the task of reflection on technology assessment; however, as the ethics example showed, it must not be faced with unfulfillable tasks of providing proof as a result of a culturally *invariant* justification. Because they have a "site in life" criteria of this kind need a basis in the factual elements of society, itself culturally *variant* – but far from being arbitrary (Grunwald 1999f). This interrelationship between acceptance and acceptability, between the factual and the counterfactual, between individual preferences and the common good, between continuity and change constitutes the field of reflection on rationality for shaping the scientific and technological future.

> ***Box 10:*** *Neglecting the relevance of factual acceptance as foundation of rationality assessments leads to "normativistic fallacies". Ethical advice, in this case, only leads to empty appeals without reaching practical relevance (Grunwald 1999f). In the global warming example, therefore, the demand to be responsible for future generations must be related to acknowledged forms of taking responsibility for the future (see above) instead of merely calling for climate protection. The forms acknowledged already used as foundation then may be developed further in order to meet new challenges.*

The recognition of the indicated preceding consensus is one of the fundamental conditions for the formation of our societies and their moral practices, something

---

[27] It is easy to imagine types of societies *without* such basic elements of ethics in the understanding mentioned. These are, instead, relying exclusively on other mechanisms of conflict resolution, for example on the principle of „survival of the fittest".

which could not simply be done away with: the underlying, partly explicit and partly implicit "social contract" may be modified on a medium time scale but it cannot be cancelled and replaced by a completely new one (on exception, perhaps, in very few revolutionary situations in history). Thus, exactly this structure of society enables its institutions to realise some forms of stable planning and decision-making over time. Exactly this structure is ensuring societal continuity through its modifications.

As an example, consider the first paragraph of the German constitution (Grundgesetz) "The dignity of Man is unimpeachable" (Die Würde des Menschen ist unantastbar). This kernel of the German constitution is normative – in the factual practices of society it is not always followed. In this way, it expresses a self-obligation of the members of this society and is, in turn, to be used to impose this obligation on citizens not following it. The normativity involved can only become relevant if it is acknowledged in main parts of society. Therefore, it must be accepted as a self-obligation and is, thus, a source for *acceptability postulates*. This part of the "social contract" of society has two essential features,

(1) It cannot be changed within short-ranged time scales. Therefore it allows continuity and stability of society;
(2) It can, however, be changed in the long run. We cannot anticipate in what way the identity of society will change in the distant future.

Therefore, such kernel elements of society show cultural variability. There is no way to prove them to be invariant with respect to culture or to history like overstressed rationality claims – the Cartesian paradigm – had in mind (compare the introductory chapter to this volume by John Grin). The variance in time, however, is small: changes - very important – do not occur in an arbitrary short-ranged way: *any modification is assessed relative to the former status and has to be justified contextually*. These contextual assessments, therefore, have to be considered against the background of the (partly decontextualised) normative framework of society – against the 'social contract'. In this way acceptability criteria as a result of rationality assessments relate new technological options to the foundations of culture and, therefore, to our past. This mechanism relates the concept of acceptability to directed incrementalism developed in section 6.3.

8   Acceptability instead of Acceptance – a Problem for Democracy?

Basing technology policy not simply on factual acceptance but on a normative basis for reasonably justified acceptability criteria raises the question as to whether there might occur a conflict between relying on normative acceptability and the idea of democracy. Is there a contradiction between relying on normative acceptability and the mechanisms of democratic societies? Is the reproach justified that there could be a tendency to exert expertocratic power? The answer to these questions is no for the following considerations:

(1) Rationality, in the way mentioned above as the source of normative acceptability criteria is (implicitly) created by people in the manifold scope of societal practices. Therefore, it is acknowledged and followed (perhaps implicitly) by people. It is not forced upon them from an external point of view or
by some experts (experts sometimes are needed to make this normativity *explicit*).[28]

(2) The way proposed in this chapter is widely accepted within our culture. Consider, for example, the case of planning a highway or a chemistry plant. There
are democratically accepted procedures for handling conflicts between the
"common good" and the individual interests of people concerned. By conducting such procedures, conflicts such as the dilemma between long-term
planning requirements and short-ranged acceptance problems can be com-
promised, mediated and processed in a way so that severe negative consequences for society are prevented (Burns and Ueberhorst 1988; Mohr 1998).
If the procedural steps are followed correctly the result has to be accepted by
the people affected. These procedures constitute the legitimisation for the results (consisting, at this level of political reasoning, of certain decisions on
technology to be followed in society).[29]

(3) The kernel of rationality in these procedures, therefore, is the insight that the
procedures must be *accepted and legitimised by democratic processes* if the
results of going through them should be accepted even if they *conflict with
interests of certain people*. The results of legitimate procedures must – and
this is part of the underlying agreement on democracy – be accepted *even if
they are not welcome. The acceptance of the procedures leads to the acceptability of their results*. This is the kernel of self-obligation within the 'social
contract', being the normative basis of democracy.[30]

In this way, it becomes possible that the preceding acceptance of the procedures
(constituted by democratic decision-making processes) leads to the acceptability
of the results. Missing or low acceptance of the results may, in this way, be overruled *in a legitimated manner*.

This mechanism, however, only works in the „normal" mode of democratic societies. The situation is changed dramatically, if legitimate decisions in technology
policy (in the sense mentioned above) lead to severe problems in society, for example in cases of complete rejection of the results (Burns and Ueberhorst 1988;
Renn and Webler 1996). Relative to criteria of pragmatic political prudence then

---

[28] The question if an expert or a layperson approach in the decision-making process should be
chosen is not dealt with at this level of consideration. Both approaches lead to analogous problems of handling the contradictions between the common good and individual interests. Only
the range of persons involved is differing.

[29] Obviously, this mechamism works only up to a certain point where, for example, dramatic
acceptance problems occur and lead to the necessity to revise the procedures (see below).

[30] This principle also governs participative procedures like public discurses (Renn and Webler
1996) or consensus conferences (Agersnap 1992). They are open with respect to the result –
and this openness must be acknowledged by the participants. The openness relates to a kind of
‚risk' in participating in such discourses: one cannot know *ex ante* if the results will be
welcome to oneself – but one has to accept the results even in the unwelcome case.

the decision may not be put through, although it has been made by legitimate procedures. In this case a societal learning process should be set up in order to review the decision-making procedures with respect to their suitability. It might be that the procedures would have to be modified. A learning process in society should result in new or modified, again legitimated procedures.

The process of modifying decision-making procedures in a legitimated manner leads to the problem how the acceptance of the modifications could be ensured. Following the approach elaborated in this paper the answer should be that these modifications should be agreed upon by going through certain "modification procedures" specially designed for this purpose. Obviously, we are running into an infinite regression ad absurdum, inventing even higher levels of procedures for modifying the procedures for modifying ... The end of such a seemingly infinite chain of procedures at higher levels can only be found in the direct acceptance of certain *basic* procedures, set into practice by an act of direct vote, like establishing the *constitution* for the society. The chain of creating legitimation comes to an end in meeting the direct vote (see to this problem also Habermas 1988; 1991, 1993).

This reflection shows that legitimate democratic procedures themselves are subject to criticisms, reviews, modifications or even replacements if they have been proven to be unsuitable to handle the challenges society is confronted with. Democratic legitimisation, therefore, can only be given in a provisional way (analogously to the "moral provisoire" of the ethics, compare Hubig 1999). Severe acceptance problems may indicate such requirements for revision (Renn and Webler 1996). The process of modifying the democratic decision-making procedures is not a short-ranged one; it is related to the underlying rationality of society and its rather slow variance. Therefore, there is no contradiction between the evolution of democracy and maintaining long-term considerations based upon democratic procedures.

In summary, there is no contradiction between the normative approach proposed and the established democratic procedures. Instead, the analysis presented fits very well to the democratic understanding of society. It points out the fact that results of democratic procedures may be legitimate even if they do not lead to complete acceptance. There is an important difference between the legitimisation of the procedures leading to acceptability of the results and the mere acceptance of the results.

9 The Role of Technology Assessment

The analysis presented above shows that technology policy – advised by technology assessment (TA) – can deal with the challenges of the present situation (between long-term planning requirements and short-ranged acceptance-orientation) by using different approaches. This leads to the question how TA may contribute to shaping the future by supporting political judgement on technology policy. [31]

At first sight, there are important clues to be found in exiting TA literature. Since the early 1980s the emphasis in TA has been shifted dramatically to the aspects of *shaping* technology and taking into account acceptance and acceptability problems. In this way, the *normative* dimension of technology policy comes to the foreground. The *question how to achieve legitimisation while one is facing (sometimes severe) acceptance problems* has developed to one of the most interesting questions in the TA debate. Recent TA approaches are taking into account the stakeholder's views. They are trying to broaden the perspectives for shaping technology and are awaiting to reach more creative and (this is in close relation to the topic of this chapter) *better accepted* technologies. This point, however, is handled differently by the various approaches, take for example Consensus Conferences (Agersnap 1992), Cooperative Discourse (Renn and Webler 1996), Constructive TA (Rip et al. 1995a) or the Interactive TA recently proposed (Grin et al. 1997). All these approaches are taking into account the question how to arrive at a legitimate technology policy very seriously. In my opinion, however, they do not answer the question 'Where does the normativity (indispensable if talking about shaping technology) come from?' sufficiently. In the following I will outline the 'rational technology assessment' approach, that deals with these issues more satisfactorily.

9.1
Rational Technology Assessment

The integrated approach "Rational Technology Assessment" recently developed in Germany (Grunwald 1998a) is based on culturalistic planning and decision theory (Grunwald 1999a, 1999c) and, accordingly, takes the perspective of shaping the future by technology instead of the prediction perspective of former TA. In this perspective the setting of goals and purposes for political, societal and technological actions becomes the most important factor instead of highlighting the unintended negative consequences and impact of technology done by former TA. Of course, possible risks and hazards of technology remain an important factor in assessing technology. Rationality assessment, additionally, requires weighing the goals and purposes against the unintended consequences. This cannot be done without explicitly considering the level of goals and purposes, of visions and *Leitbilder*. From objectives and purposes (set for technology developments as well as for societal progress) orientational knowledge for actual decisions can be achieved by *backcasting* (Grunwald 1999a). The normativity which is indispensable for giving "sustainable" recommendations (in the sense of reliable and relatively stable) for technology policy or to assess visions with respect to rationality standards has to be reconstructed in the way proposed above (sections 5, 7).

---

[31] It seems to be impossible to review the various types of TA and their relation to the main topic of this paper completely and in detail. Therefore, the considerations are focussed on the explanation of my own approach instead of discussing its relations to other approaches (except some rather brief comparisons).

> **Box 11:** *The sustainability discourse should, in this way, be established as a permanent social, political, scientific and public reflection on nature, environment and the future of society. Climate policy, as an example, can never be finished in the sense of "goal attained". Instead, permanent monitoring, observing, modeling, simulating, deciding on politicial measures etc. is required – an excellent example for the model of "directed incrementalism" and flexible planning (section 6).*

Therefore a rational TA on this basis makes it possible to formulate well-founded long-term and reliable perspectives for science and technology policy (Gethmann 1998). Its primary aim should be to make it possible to cope rationally, efficiently and productively with foreseeable uncertainties in decision-making processes and conflicts relating to science and technology. *Rational* ways of coping with conflicts relating to science and technology are a substantial pre-condition for long-term reliable policies for science and technology. For this situation the assessment of the consequences of science and technology should, wherever possible, take place to the *early stages* of development, where social control measures can have effect without serious economic consequences (Rip et al. 1995b, p 5; Hoppe and Grin 1995, p 6). This includes an anticipatory search for alternative decisions and preventive measures to avoid ruinous investments or blind alleys in research and technology policies. In this way continuous reflection accompanying scientific and technological developments (including ethical reflection in cases of moral conflicts involved, Grunwald 1996a; 1999d; 1999f) increases the efficiency in the utilisation of social resources and allows to evaluate short-term developments against the background of long-term and relatively stable social constellations and structures (section 6.3, directed incrementalism).

9.2
Rationality and Participation

Hitherto, the question was not discussed what groups of persons should be included in the decision-making process. The rejection of purely acceptance-oriented approaches, however, allows us to draw some conclusions (for an application to some aspects of shaping the information society compare Grunwald 1999g).

The most important point is that participation is not automatically related to acceptance-orientation. Instead of allowing merely existing acceptance of stakeholders to govern technology policy decisions, it is more important to consider the perspective of setting goals and aspirations for political, societal and technological action according to the underlying rationality standards of society should become more important. For participative TA, this means that involving the stakeholders should not be restricted to taking into account their acceptance behaviour. Participative TA should – and some recent work on TA seems to point in this direction (Grin et al. 1997) – produce and process acceptability criteria instead of striving to achieve acceptance. In this sense, participative TA should address the acceptance problems neither by means of an adaptive nor by a shaping approach but should set up *societal learning processes* by explicitly taking into account the level of goals and purposes mentioned above (Grunwald 1999b). Creative solutions through interaction between a variety of views on what should be done (Grin and van de Graaf 1996; Grin et al. 1997) should be the result of such learning processes.

Generating acceptance or transforming acceptance into decisions may be suitable for very special situations (Agersnap 1992; Renn and Webler 1996); but in many cases, TA must go far beyond such simple schemes. Participation, in general, should lead to a common understanding of what is to be done and of the criteria to be applied for establishing the appropriate plan of action also in the absence of complete, immediate acceptance. It should involve collective learning instead of merely mediating or compromising to attain technology acceptance (Jelsma 1995; Wynne 1995).

Learning implies argumentation: questioning one's own position and the disposition to be able to change this position in certain argumentation situations (Jelsma 1995, p 152ff.). It means that politicians learn from laypersons, laypersons from experts, experts from politicians and – very important - the other way round. These brief remarks show the advantage of the approach proposed as well as its problems. The advantage is that the "best" solution available (section 5) is looked for, agreed upon and can be sustained over time (section 7), perhaps in a modified way due to changing contextual conditions (section 6.3):

> If practised properly, TA could play an important role in reaching congruent meanings, which could then serve as the basis for joint action undertaken by different types of actors (Grin and van der Graaf 1996, p 96).

The problem is that arguing and learning are difficult procedures, much more difficult than relying simply on acceptance or – the counterposition – basing technology policy simply on authoritarian dictate. Especially, learning requires certain social pre-conditions to be fulfilled as well as a cognitive disposition to accept the better argument. And, the importance of uncovering underlying assumptions of the arguments used, seems to be evident (see the paper of Reuzel and van der Wilt and the final chapter of this book). However, it should be an attractive „vision" for the future to evolve towards a "learning society" in this sense.[32]

Furthermore, the question remains who shall participate in decision-making procedures and in processes of practical reasoning preceding them. This question is the same as is discussed in established participative approaches (Agersnap 1992; Renn and Webler 1996); it arises independent from whether acceptance or acceptability shall be generated. The only answer at this level of generality is to point to the need for contextualising the attribute "valid for everyone" (sections 5, 7). The degree of participation and its particular realisation is dependent on the context and includes the particular problem under discussion, the decisions to be made, the measures in question and the particular cultural background of that

---

[32] Learning itself is related to normative issues (Wynne 1995). The normativity involved has to be justified again by recurring to the underlying cultural rationality standards in order to avoid authoritarian biasses.

society, especially its political traditions and established and acknowledged decision-making mechanisms (Grunwald 1999g).

The range of persons who should be involved in participatory assessments
should depend on the degree to which they are affected or concerned. It is essential that groups that are to be involved *have a chance* to agree on the technology in
question. Acceptance items, however, would be over-stressed (and lead to the
unwelcome impact discussed in section 4) if it were expected that the people involved must all *in fact* agree. Their agreement may be *presumed for reasons of
cultural rationality* if the technology decision is based on acknowledged and legitimate procedures (possibly including „entry points" for expressing protest and
opposition) (section 8). Generating democratically legitimated decisions by participation implies, vice versa, that participative procedures must be embedded in
an acknowledged framework of decision-making. Otherwise, participation would
not avoid legitimisation problems (Grunwald 1999g).

Experts are always experts of "something", mostly of some very specialised
area of interest. Experts of the future are not available for principal reasons (section 5.3). Shaping the future is the task of society as a whole; it cannot be dele-
gated to experts. The only method that can be legitimated in a democratic culture
is to delegate some "shaping capability" or "shaping power" to the political system, legitimated by democratic (and accepted!) procedures. This delegation in the
representative democracy, however, is a borrowed one: it is borrowed from people. In the case of situations or problem types where the shaping methods established by the political system do not work, the sovereign (the citizens independent
from whether they are laypersons or experts) is right to reverse this delegation and
to change the decision-making procedures (see above, section 8).

9.3
Rationality and Vision Assessment

Rational TA should include some kind of vision assessment because exactly this
type of assessment is essential for ensuring some long-term planning items within
a changing society (section 5.2). Visions *per definitionem* are to give orientation
for the long run: if they were changing rapidly they would loose their power for
giving orientation. What can be learnt from the rationality considerations given
above for such vision assessments? Some aspects shall be mentioned briefly in the
following:

Vision Assessment and Normativity

Visions guiding our plans and ideas of shaping the future are, at least in important
aspects, *normative*. They include ideas about how we wish to live in future. To
reflect and assess such ideas is a highly interesting challenge for TA as well as for
ethics (Grunwald 1996a; 1999f). The most difficult task is to make explicit the
foundation of the normative background used to judge visions because technology
decisions based on such visions may burden some people or groups of society

with the consequences of that decision. If acceptance amongst the latter fails, the key question is whether the acceptability of such decisions can be made transparent. Convincing these people and groups of the acceptability implies arguing why and to what extent the commitment to the "common good" (inherent to the underlying social contract) justifies to burden them with a negative impact and disadvantages.

Acceptance according to individual preferences is mostly the result of a rather short-sighted deliberation process. Long-term visions, however, could in fact influence acceptance by parts of the population on the basis of acknowledged rationality standards (sections 5, 7). The way sometimes suggested to improve acceptance-driven technology policy by putting more effort in providing information (Mohr 1998) about future scenarios is restricted in range. Often not missing knowledge about new technologies but *normative* objections are the main reason for rejecting new technologies (biosciences, reproductive medicine). In such cases opponents cannot be convinced by providing (descriptive) information; instead, the normative premises of their objection must be analysed and checked with respect to their generalisability.

Contextualisation

Visions have to be assessed *contextually*. Their rationality cannot be evaluated merely against situation-invariant standards. The standards for rationality assessments are themselves culture-dependent (section 5). However, there are still strong claims for an invariance with respect to *individual persons* (in regarding the "common good", section 8). Furthermore, TA cannot be dissolved into pure contextualisation. If one wants to learn from past experiences with TA, one has to establish a certain kind of *decontextualisation*: learning implies the transfer of some contextually bound issues into a *different* context. The various types of contexts – at least in the present society and its near past - are interrelated by the underlying rationality of society (the 'social contract'). This relation allows some learning (section 5.4) and a directed incrementalism in shaping the future (section 6.3); complete contextualisation would not leave any chance for bridging the gaps between different contexts.

Broadening the Perspective

The assessment of visions must respect the pluriformity of the views of people affected. This statement does not pre-decide the question if there should be a participatory procedure, an inquiry or an expert approach taking into account those views (if they are well-known). Technology policy based on visions must be acceptable to the people concerned. However, it is not reasonable to try to reach completeness with respect to the views of people. Instead of such *strong universalism* it seems to be more suitable to start with *prudent and pragmatically correct relevance decisions.*[33] Broadening the perspective alone (Thompson and Schwarz 1990) seems to be necessary *but not sufficient* to arrive at acceptable technology policies. The mere pluriformity must be transcended by assessing the "validity for everyone" to create generalisable statements which then could be used in democratic procedures to reach legitimate and transparent decisions.

Interdisciplinarity

Vision assessment must be multidisplinary to achieve a concurrent view of the various disciplines involved in that particular problem (compare the paper of Michael Decker). Because this point seems to be agreed upon in TA it requires no further explanation.

Time Scales

Vision assessments have to take into account the dependencies of social ideas on time very carefully. Consider, for example, the case that a vision loses its orientational power or that it seems to lead into false directions. Such cases must (a) be foreseen as early as possible in order to avoid investments in blind alley developments and (b) be taken into account in the initial agenda. To achieve this it is very important to implement points in the decision procedures and development paths where modifications or even complete abandoning of that vision should be possible (according to the model of directed incrementalism, section 6.3).

TA as a Process

The directed incrementalism as a model of shaping technology relates to the concept of TA as a *process* instead of TA as a *result* (Paschen and Petermann 1991; van Eijndhoven 1997): vision assessment cannot be done within single reflection processes leading to fixed results. Instead, it must consist of a permanent assessment and reflection process, accompanying the technological development. This accompanying reflection has to investigate the relation of that vision to the (perhaps changing) societal contexts and to the (relatively stable) underlying 'social contract'. This reflection allows balancing short-ranged flexibility and long-term stability requirements. It constitutes a (normative) shaping process, not a (descriptive) forecasting process (Grunwald and Langenbach 1998).[34]

---

[33] Within the CI example (see the paper of Rob Reuzel and Gert van der Wilt) it was not required to take into account *all* possible views but to take notice of the views of the people really concerned.

[34] TA as an accompagnying reflection process requires certain guidelines and procedures for ensuring its quality and reliability (compare van Eijndhoven 1997, p 283f).

Vision Assessment and Cultural Rationality

The assessment of visions should be based on investigations of the underlying factually accepted normativity of society. Then questions like "Is the vision under consideration compatible with this normativity?", "Are there learning and adaptation processes required in following this vision?" could be answered. In general, there wouldn't be any chance for proceeding further with this vision if the assessment shows that it is too far away from the actual state of the "social contract".

Limits of Procedural Legitimisation

Procedural legitimisation by established mechanisms is subject to some limitations (section 9). Questions to be answered to make the range of the concept of acceptability more transparent, are: "Where are justified criteria for maintaining a development with low acceptance?", "How should be distinguished between short-ranged acceptance problems (due, perhaps, to a contingent event like an accident) which could be overruled by proceeding with the initial agenda (perhaps, with modifications) on the one hand and severe acceptance problems on the other, pointing to serious incompatabilities between the underlying normativity of society and the technology under consideration?", "How far does the legitimisation of democratic procedures reach in cases of non-accepted technologies?", "Where is the „break-even point" beyond which it would be more prudent to abandon a technology though its genesis has been completely legitimated?". Answering these questions requires more detailed analysis of the process of technology development and the democratic decision processes and shall be left to further work.

Technology Assessment and the Theory of Society

In the last consequence, vision assements are related to the image we have of the present state of our society, of its underlying social contract and the acceptance of procedures for generating legitimated decision results. Therefore, technology assessment is related to the *scientific* image we have of society – this is essentially constituted, with respect to descriptive features, by the social sciences, and, with respect to the normative issues involved, by the philosophical ethics (Grunwald 1999f). In this way, here is a need for more cooperation between social sciences and ethics in order to improve our understanding of the relation of normativity and acceptance.

Vision assessment in this way may consist of a consultation on the policy process as well as deliberation within it. Rational TA points to the importance of legitimate decision procedures. It allows widening the principle of representation by means of participation – again, of course, only in a legitimated version. In this approach, TA should be an accompanying activity or a reflective instance delivering deliberative facilities to contribute to shaping technology in and by society.

10 Conclusion

The dilemma central to this chapter seems to result from a severe difference between presupposed rationality models:

- the acceptance-orientation leading to short-ranged policies without any direction in the long run restricts itself to descriptive understanding of TA and leaving the normativity of shaping the future to the individual preferences. Thus, it favours contextualisation and bottom-up procedures in defining goals and assessing visions;
- long-term requirements and their obligations for people are highlighted within the rationality standard of High Modernity. This approach favours situations-invariances and decontextualisation (because of some metaphysical or transcendental presumptions) and uses top-down procedures for decision-making and assessments.

The pragmatic rationality concept used as kernel of the approach presented (sections 5, 7) should allow us to overcome the gap between these two approaches and to follow long-term paths and visions into the future as well as to include short-ranged acceptance problems in practical reasoning about the future.

References

Agersnap T (1992) Consensus Conferences for Technology Assessment. In: Technology & Democracy. Proceedings of the 3th European Conference on Technology Assessment, Copenhagen, pp 45-54

Arrow K (1963) Social Choice and Individual Values. Yale University Press, London

Banse G, Friedrich K (1996) Sozialorientierte Technikgestaltung – Realiät oder Ilusion? Dilemmata eines Ansatzes. In: Banse G, Friedrich, K (eds) Technik zwischen Erkenntnis und Gestaltung. Edition Sigma, Berlin, pp 141-164

Bayerische Rück (1992, ed.) Risiko ist ein Konstrukt. Knesebeck

Bechmann G (1991) Folgen, Adressaten, Institutionalisierungs- und Rationalitätsmuster: Einige Dilemmata der Technikfolgenabschätzung. In: Petermann Th (ed) Technikfolgen-Abschätzung als Technikforschung und Politikberatung. Campus, Frankfurt, pp 43-72

Bechmann G (1994) Frühwarnung - die Achillesferse der TA? In: Grunwald A, Sax H (eds) Technikbeurteilung in der Raumfahrt. Edition Sigma, Berlin, pp 88-100

Bechmann G, Coenen R, Gloede F (1994) Umweltpolitische Prioritätensetzung. Verständigungsprozesse zwischen Wissenschaft, Politik und Gesellschaft. Metzler-Poeschel, Stuttgart

Beck U (1992) Risk Society. Towards a new Modernity. Sage, London

Bijker W, Law J (1994, eds) Shaping Technology Building Society. MIT Press

Bijker WE, Hughes TP, Pinch TJ (1987, eds) The Social Construction of Technological Systems. New Directions in the Sociology and History of Technological Systems. Cambridge (Mass.)/London

Boulding K (1964) Review of a Strategy of Decision. American Sociological Review 29, pp 921-942

Brauch HG (1996) Klimapolitik. Springer, Berlin Heidelberg New York

Braybrooke D, Lindblom CE (1963) A Strategy of Decision. New York

Bröchler S (1998): A New Framework for Technology Assessment: Aspects of the Concept of Innovation-Oriented TA. In: Proceedings of the 2nd International Conference on Technology Policy and Innovation, Lisbon 1998, pp 25.1.1-25.1.8

Brown J (1988) Rationality. London/New York

Burns TR, Ueberhorst R (1988) Creative Democracy. Systematic Conflict Resolution and Policymaking in a World of High Science and Technology. Praeger , New York

Camhis M (1979) Planning Theory and Philosophy. London

Collingridge D (1980) The Social Control of Technology. New York

Dierkes M, Hoffmann U, Marz L (1992) Leitbild und Technik. Zur Entstehung und Steuerung technischer Innovationen. Berlin

Gethmann CF (1982) Proto-Ethik. Untersuchungen zur formalen Pragmatik von Rechtfertigungsdiskursen. In: Elwein Th, Stachowiak H (eds) Bedürfnisse, Werte und Normen im Wandel, Vol 1. München, pp 113-143

Gethmann CF (1992) Universelle praktische Geltungsansprüche. Zur kulturellen Genese moralischer Normen. In: Janich P (ed) Entwicklungen der methodischen Philosophie. Suhrkamp, Frankfurt, pp 146-179

Gethmann CF (1996) Rationalität. In: Mittelstraß J (ed) Enzyklopädie Philosophie und Wissenschaftstheorie, Vol 3. Metzeler, Stuttgart, pp 468-481

Gethmann, CF (1998): Rationale Technikfolgenbeurteilung. In: Grunwald A (ed.) Rationale Technikfolgenbeurteilung. Konzeption und methodische Grundlagen. Springer, Berlin Heidelberg New York, pp 1-11

Gottschalk N, Elstner M (1997) Technik und Politik. Überlegungen zu einer innovativen Technikgestaltung. In: Elstner M (ed) Gentechnik, Ethik und Gesellschaft. Heidelberg, pp 143-180

Grin J, Graaf H van den, Hoppe R (1997) Technology Assessment through Interaction, Amsterdam

Grin J, van de Graaf H (1996) Technology Assessment as Learning. Science, Technology & Human Values Vol 21, Sage Publ., pp 72-99

Grunwald A (1994) Wissenschaftstheoretische Anmerkungen zur Technikfolgenabschätzung: Prognose- und Quantifizierungsproblematik. Journal for the General Philosophy of Science 25/1, pp 51-70.

Grunwald A (1996a) Die Bewältigung von Technikkonflikten. Theoretische Möglichkeit und praktische Relevanz einer Ethik der Technik in der Moderne. Zeitschrift für philosophische Forschung 51, pp 437-452

Grunwald A (1996b) Sozialverträgliche Technikgestaltung. Kritik des deskriptivistischen Verständnisses (Graue Reihe No 3). Europäische Akademie, Bad Neuenahr-Ahrweiler

Grunwald A (1998a, ed) Rationale Technikfolgenbeurteilung. Konzeption und methodische Grundlagen. Springer, Berlin Heidelberg New York

Grunwald A (1998b) Rationale Gestaltung der technischen Zukunft. In: Grunwald A (ed.) Rationale Technikfolgenbeurteilung. Methodische Grundlagen und Verfahren. Springer, Berlin Heidelberg New York, pp 29-54

Grunwald A (1998c) Technisches Handeln und seine Resultate. Prolegomena zu einer kulturalistischen Technikphilosophie. In: Hartmann D, Janich P (eds) Die kulturalistische Wende. Suhrkamp, Frankfurt, pp 178 - 224

Grunwald A (1999a) Handeln und Planen. Philosophische Planungstheorie als handlungstheoretische Rekonstruktion. Bouvier, Bonn

Grunwald A (1999b) Limitations of an Acceptance-Oriented Technology Policy? The IPTS Report 33 (to appear)

Grunwald A (1999c) Rationality in Shaping Technology? In: Hronszky I et al. (eds) Studies on the policy of Science, Technology and Environmental Issues. Budapest: MTA Szociologiai Intezete, (to appear)

Grunwald A (1999d) Technology Assessment or Ethics of Technology? Reflections on Technology Development between Social Sciences and Philosophy. Ethical Perspectives (to appear)

Grunwald A (1999e) Against Over-Estimating the Role of Ethics in Technology. Science and Engineering Ethics (submitted)

Grunwald A (1999f) Ethische Grenzen der Technik? In: Grunwald A, Saupe S (eds) Ethik in der Technikgestaltung. Praktische Relevanz und Legitimation. Springer, Heidelberg Berlin New York, pp 221-252

Grunwald A (1999g) Rationality and Participation. On the Role of Technology Assessment in the Knowledge Society. In: Banse G, Langenbach CJ, Machleidt P (eds) From an Information Society to a knowledge Society. Heidelberg: Springer (in preparation)

Grunwald A, Langenbach C (1998) Die Prognose von Technikfolgen. Methodische Grundlagen und Verfahren. In: Grunwald A (ed.) Rationale Technikfolgenbeurteilung. Methodische Grundlagen und Verfahren. Springer, Berlin Heidelberg New York, pp 93-131

Gutmann M (1998) Der Begriff der Kultur. In: Hartmann D, Janich P (eds) Die kulturalistische Wende. Suhrkamp, Frankfurt, pp 269-332

Habermas J (1987) Zwecktätigkeit und Verständigung. Ein pragmatischer Begriff der Rationalität. In: Stachowiak H (Hrsg) Pragmatik. Handbuch pragmatischen Denkens. Hamburg, pp 32-59

Habermas J (1988a) Theorie des kommunikativen Handelns. Suhrkamp, Frankfurt

Habermas J (1988b) Nachmetaphysisches Denken. Suhrkamp, Frankfurt

Habermas J (1991) Erläuterungen zur Diskursethik. Suhrkamp, Frankfurt

Habermas, J (1993): Faktizität und Geltung. Suhrkamp, Frankfurt

Harig H, Langenbach C (1999, eds) Neue Materialien für innovative Produkte. Springer, Heidelberg Berlin New York

Hartmann D, Janich P (1996) Methodischer Kulturalismus. In: Hartmann D, Janich P (eds) Methodischer Kulturalismus. Zwischen Naturalismus und Postmoderne. Suhrkamp, Frankfurt

Hartmann D, Janich P (1998, eds): Die kulturalistische Wende. Suhrkamp, Frankfurt

Hempel C G (1965) Aspects of Scientific Explanation and other Essays in the Philosophy of Science. New York, London.

Hohmeyer O, Hüsing B, Maßfeller S, Reiß T (1994) Internationale Regulierung der Gentechnik. Heidelberg

Hoppe R, Grin J (1995) Technology Assessment for Participation: Experiences and Lessons. Industrial & Environmental Crisis Quarterly 9, pp 3-12

Jelsma J (1995) Learning about Learning in the Development of Biotechnology. In Rip, A., Misa, Th., Schot, J., 1995 (eds), Managing Technology in Society. London, pp 141-166

Jonas H (1979/1984) The Imperative of Responsibility, Chicago, 1984 (German: Das Prinzip Verantwortung. Versuch einer Ethik für die technologische Zivilisation. Frankfurt 1979)

Jungermann H, Slovic P (1993) Charakteristika individueller Risikowahrnehmung. In: Krohn W, Krücken G. Riskante Technologien: Reflexion und Regulation. Suhrkamp, Frankfurt, pp 79-100

Kern L, Nida-Rümelin J (1994) Logik kollektiver Entscheidungen. Oldenbourg, München

Kettner, M. (1999): Neue Wege der Diskursethik. In: Grunwald A, Saupe S (eds) Ethik in der Technikgestaltung. Springer , Berlin, Heidelberg, New York, pp 153-196

Kuik O, Verbruggen H (1992, eds) In search of indicators of sustainable development. Dordrecht

Ladd J (1975) The Ethics of Participation. Nomos 16, pp 98-125

Law J, Bijker W (1994): Postscript: Technology, Stability and Social Theory. In: Bijker W, Law J (eds) Shaping Technology Building Society. MIT Press, pp 290-308

Lorenzen P (1987) Lehrbuch der konstruktiven Wissenschaftstheorie. Bibliographisches Institut, Mannheim

Lorenzen P, Schwemmer O (1973) Konstruktive Logik, Ethik und Wissenschaftstheorie. Bibliographisches Institut, Mannheim

Lübbe H (1997) Modernisierung und Folgelasten. Springer, Berlin, Heidelberg, New York

Luhmann N (1990) Die Wissenschaft der Gesellschaft. Suhrkamp, Frankfurt

Mohr H (1998) Technikfolgenabschätzung in Theorie und Praxis. Springer, Berlin, Heidelberg, New York

Moore G (1904) Principia Ethica. Cambridge

Paschen H, Petermann Th (1991) Technikfolgenabschätzung - ein strategisches Rahmenkonzept für die Analyse und Bewertung von Technikfolgen. In: Petermann Th (ed) Technikfolgen-Abschätzung als Technikforschung und Politikberatung. Campus, Frankfurt, pp 19-42

Renn O (1992) Concepts of Risk: A Classification. In: Krimsky S, Golding D (eds) Social Theories of Risk. Westport

Renn O (1996) Kann man die technische Zukunft voraussagen? In: Pinkau K, Stahlberg C (eds) Technologiepolitik in demokratischen Gesellschaften. Stuttgart, pp. 23-51

Renn O, Webler Th (1996) Der kooperative Diskurs: Grundkonzeption und Fallbeispiel. Analyse&Kritik 18, pp 175-207

Rescher N (1988) Rationality. Cambridge

Rip A, Misa T, Schot J (1995a) (eds) Managing Technology in Society. London

Rip A, Misa T, Schot J, (1995b) Constructive Technoology Assessment. A New Paradigm for Managing Technology in Society. In: Rip et al. (1995a) (eds), Managing Technology in Society. London, pp 1-14

Schumpeter J (1934) Theorie der wirtschaftlichen Entwicklung. Duncker & Humblot, Berlin 1993

Schurz G (1988) (ed.) Erklären und Verstehen in den Wissenschaften. Oldenbourg, München

Schurz G (1995) Grenzen rationaler Ethikbegründung. Das Sein-Sollen-Problem aus moderner Sicht. Ethik und Sozialwissenschaften 6 (1995), pp 163-177

Schwarz M, Thompson M (1990) Divided We Stand. Harvester Wheatsheaf Press, Hassocks

Shrader-Frechette KS (1991) Risk and Rationality. Philosophical Foundations for Populist Reforms. Berkeley

TAB, Büro für Technikfolgenabschätzung des Deutschen Bundestages (1997, ed) Technikakzeptanz und Kontroversen über Technik – Ambivalenz und Widersprüche. TAB-Arbeitsbericht Nr. 54

Todt O (1997) The Role of Controversy in Engineering Design. Futures 29(2), pp 177-190

Todt O, Lujan L (1998) Social Planning of Technology. The IPTS Report No. 26, pp 30-34

Toulmin S (1989) Cosmopolis. The Hidden Agenda of Modernity. Macmillan, New York

van Eijndhoven J (1997) Technology Assessment: Product or Process? Technological Forecasting and Social Change 54, pp 269-286

van Gunsteren HR (1976) The Quest for Control: A critique of the raional-central-rule approach in public affairs. John Wiley, London

VDI, Verein Deutscher Ingenieure (1997, ed) Pragmatische Maßnahmen zur Förderung der Technikaufgeschlossenheit in Deutschland. Düsseldorf

Wagner-Döbler R (1989) Das Dilemma der Technikkontrolle. Edition Sigma, Berlin

Wynne, B. (1995): Technology Assessment and Reflexive Social Learning: Observations from the Risk Field. In: Rip, A., Misa, Th., Schot, J. (eds), Managing Technology in Society. London, pp 19-36

Replacing Human Beings by Robots.
How to Tackle that Perspective by Technology Assessment?

Michael Decker

1   Introduction

'May I help you?', a friendly voice asks politely as one enters the main entrance of the hospital. The question was asked by a five feet tall robot which focuses on you with two digital cameras. Depending on your wishes the robot leads you to the emergency admission, to the patient you want to visit, or to the administration of the hospital. On the way to the emergency admission one meets robots that serve meals and feed patients, as well as autonomous wheelchairs that transport patients to different treatment or examination rooms. After physicians have finished their examination they ask a software system to evaluate their diagnosis. Robots also dominate operating rooms. They assist in minimal invasive surgery, bore holes with high precision, cut tissue and sew the wound again, while the surgeon is sitting in a high-tech chair like a pilot and operates the robot with a joystick. In the intensive care unit all patients are monitored by robots that record blood pressure, pulse, oxygen level in the blood, etc., and are fed with information on the treatments the patient receives. From these data the robot can generate suggestions for further treatments.

If one believes the prognosis contained in the Delphi Report (BMBF 1997), an up-to-date German study into future trends in science and technology, based on the questioning of over 2000 experts, the hospital described above could become true at the latest in the year 2025. In fact, prototypes of most artificial agents occurring in this scenario already exist.

Following Rob Reuzel and Gert Jan van der Wilt in their definition of "vision" (this volume), I would argue this vision of "autonomous robots in future health care" describes, how future hospitals should look like according to researchers in the field of robotics.[1] In order to construct a more balanced vision (cp. the central question formulated in the introductory chapter by John Grin), it is necessary to combine the robotics perspective with others.

---

[1] From my point of view the individual researcher on robotics will not take the vision of the future hospital as a whole as his guiding vision (cp. Peter Mambrey 4.2 level two and three). He would only focus on the artefact he wants to develop in the particular context. The combination of these single visions to the vision of a future hospital is already looked upon as a task of Pre-TA.

According to the definition of guiding visions "as a medium that establishes 'structural interconnections' between different and largely autonomous social subsystems" given by Peter Mambrey (in this volume), this vision is still at the stage of "individual cognitive actuation" of the social subsystem of researchers in the field of robotics. There are no other social subsystems developing their own vision so far, and up to now there have been no traces of public awareness . The guiding function as a collective projection (Mambrey) should be started now and this is interpreted as the first step of technology assessment in this contribution.

It is widely agreed that Technology Assessment (TA) ideally should start in early stages of technology development (Grunwald 1998; Paschen and Petermann 1991; Peter Mambrey in this volume). In this paper I would like to suggest that the assembling of an interdisciplinary expert groups a good starting point for such assessment.

Armin Grunwald (1998; cp. 9.3 in this volume for more details) advocates so-called Rational TA as a way of scrutinising visions and developing visions that are acceptable to society. Following Grunwald in his approach, I, in this contribution, will address the following aspects of Rational TA, which are particularly relevant for assessing technologies at an early stage:

- *Broadening the perspective and inter-disciplinary scope*
  In the approach proposed in this contribution, experts from relevant scientific disciplines are called in to add their perspectives to an initial concept of the technology.
- *Contextualisations*
  I will refer to cases in the health care area.
- *Normativity*
  Recommendations as to 'what should be done' are explicitly given and based on transparent argumentation.
- *TA as a process*
  The results of my approach should be seen as a first step in an ongoing TA-process. Therefore, an unbroken chain of argument will be developed and represented. In this way, the results are subjected for criticism and further discussion.

Due to the fact that, in the case of robots in a hospital setting, the perspective of developers is the first one existing,[2] it is sensible to use it as a basis for the interdisciplinary discussion of the expert group. The expert group should start with structuring the field of interest for three reasons. Firstly, one can derive from this structure (a) which views will be important and for which reasons, and (b) which disciplines should be involved in the discussion accordingly. Secondly, the discussion itself can be based on such a structure. Thirdly, the structure can be useful for subsequent steps in the TA-process, for example, the identification of stakeholders.

Therefore, in section 2 a proposal is made for structuring the scientific field of autonomous robots, in order to prepare it for Technology Assessment. The way outlined here starts, as mentioned above, from the technical perspective, from

---

[2] In other cases often a problem-oriented perspective is the first one existing.

which autonomous robots are developed to replace human functions. In the line of this thesis, one can define different levels of replaceability relative to the functioning of the robot in the overall system of human *and* robot. Apart from these levels, one can discuss various areas of human replaceability. In section 3, this two-dimensional structure is evaluated by reflection on three examples from the health care area.

In section 4, it is suggested how TA in the field robotics can be proceed on the basis of the structure developed so far. Experts, one or two from each discipline involved, are invited to an interdisciplinary debate, in which they have to develop a perspective on the vision "autonomous robots in future health care". Subsequent combination of these perspectives to one "more complete and balanced vision" will take place according to several rules of discussion, so that transparency of argumentation is ensured. In this combination process assumptions underlying individual perspectives will be uncovered and criticised from other disciplines. The results of this interdisciplinary discussion will be published in a memorandum that indicates areas where no autonomous robots should be put into action and for what reasons. This is an important contribution to Technology Assessment of "autonomous robots in future health care".

Sometimes, this expert-memorandum can already[3] be addressed to the legislative body; for example, acceptable maximum error ratings can then be stipulated by law. Alternatively, the memorandum can be sent to the executive body, so that the provision of funds for the promotion of research can be concentrated on those areas within which artificial agents appear to be most valuable. The research community itself will also welcome these results in order to be able to focus on areas most suitable to furthering the replacement of human beings by artificial agents. Finally, it could be a basis for the participation of laypersons.[4] Here, the expert group should either directly participate in a discussion with laypersons, or be prepared to subject their memoranda to a dialectic and iterative process of deliberation in which experts and laypersons comment on one another's documents, rather than their personal appearance. Involving laypersons may fulfil two crucial functions. First, it broadens the normative scope of the assessment and thus optimises acceptability in the sense discussed in the chapter by Armin Grunwald. Second, co-operation of laypersons, e.g. users, may be important for improving the effectiveness of the technology. The important thing is to uncover or construct a shared perspective on the technology. This normatively and pragmatically validated vision could then also be translated into more comprehensive recommendations, and addressed to the recipients already mentioned above.

---

[3] There may be contexts, where obviously no conflicts occur and therefore a further discussion with other experts or laypersons need not to be pursued.

[4] Which should be aimed at in some contexts of robotics

2 Structuring the Field for a TA-Process. Which Criteria seem to be Relevant?

2.1
Thesis: It is the Purpose of an Artificial Agent to Replace Human Beings

When developing an artificial agent for specific tasks, it is first of all necessary to clarify the scope of the functions required. For this purpose two main standpoints must be considered, i.e. that of those demanding a new functional agent and that of those who have hitherto performed the tasks involved. (If it is a question of new functions not hitherto performed by anyone, it only remains to consider the demands of the potential purchaser)[5]. From these considerations, it will be possible to compile a catalogue of requirements to be met by the artificial agent. Usually, this catalogue will define these requirements according to those that are absolutely necessary (minimum requirements), and those that should also be aimed at, but are not essential (Janich 1993).

Once the designer of the robot has produced a prototype, this prototype will be evaluated against the criteria set out in the catalogue. Of course, it must achieve the minimum aims, but in order to be competitive on the market it must also meet as many of the additional desirable requirements as possible. To enable reliable evaluation of the agent's functional capabilities, the requirements should be translated into applicable criteria, i.e. operationalised. Typically, one would define intervals corresponding to measurable performance indicators.

The purpose-oriented approach implies that the artificial agent will fulfill functions that would otherwise have to be fulfilled by a human being; it is clear that the purpose does not alter simply because the functions are performed by an artificial agent. On the contrary, the artificial agent is merely a means to an end in the fulfilment of a given purpose[6]. If the development of an artificial agent for a specific aim is not successful, then humans must take another way in order to achieve their aim. It does not matter how as long as the criteria set before are met.

To illustrate this, let us contemplate a robot that is able to operate the camera during a minimal invasive surgical operation (Hirzinger 1996). Here, the surgeon makes the following demands on his camera assistant (minimum requirements): a) hold the camera still without wobbling, b) move the camera without translating any movement to the opening of the incision, c) act on commands to change the viewing angle. To construct and evaluate a robot for these tasks, the requirements must be translated into performance indicators with respect to a) the frequency range and amplitude of camera wobble allowing for clearly recognisable pictures,

---

[5] Please keep in mind that this thesis is rooted in the technical perspective of engineers. The demands of other stakeholders are less important in the context of developing a well functioning artefact.

[6] According to (Dewey 1922, Rohbeck 1993) it is also possible that new aims can arrive from the existence of available means. In the case discussed here this seems to be less important.

b) the acceptable degree of movement translated to the opening of the incision, and c) the acceptable deviation of the actual range of vision from the range of vision specified. If it is possible to build a robot meeting these criteria, then it may be used as a camera assistant.

It is obvious that these functions need not be performed in exactly the same manner as a human camera assistant would have done. Whereas, for example, the assistant doctor would probably use one of his hands to hold the camera mounting firmly at the opening of the incision to prevent translation movement, the algorithm of the robot control mechanism would automatically preclude translation movement in this area.

Thus the purpose of an artificial agent is in this way to take over tasks previously done by humans. Now, the functional performance of the robot within the overall man/machine system can be sub-divided as follows.

2.2
Overall capability levels

As shown in the previous paragraph, the human being employs artificial agents to achieve certain aims. In order to evaluate the overall performance of the resulting man/machine system, a standard functional capability scale is needed, whereby the standard functional capabilities of a human could be taken for a benchmark . Clearly, in each case, it is necessary to relate the standard capabilities to a specific task.[7] The criteria on the basis of which the capabilities of the artificial agent will later be evaluated should correspond to this.

Here again the question arises as to how such standard capabilities of the human body should be defined. Obviously, the physical performance of a top-class sportsman would not be appropriate. Instead, the standard should be based on a yet-to-be-defined "average performance", which, though not yet precisely articulated, may be presumed to exist. On the basis of this assumption three types of capability can be distinguished (Decker 1997)

1. Some human beings are unable to reach standard capability levels due to a physical handicap, disease or injury. In such cases, artificial agents may be employed to compensate for these deficiencies. Then the artificial agent involves a RESTITUTION of the capability. "Intelligent" prostheses belong to this category, which take over the functions of certain organ or limbs, autonomous robots for getting and bringing services, or wheelchairs that are able to lift handicapped people into or out of the bed.
2. With respect to other functions, human beings are replaced by robots. This type of application may be referred to as SUBSTITUTION. In these cases, standard human capabilities are defined in relation to the specific functions required, in order to provide criteria for the evaluation of the artificial agent. An example of this type is the "camera assistant" described above.

---

[7] Cp. the catalogue of requirements aforementioned.

3. If the overall capability of a combined man/machine system exceeds that of the corresponding human capability, this application constitutes an EXPANSION of the capability. Take, for example, a large database, which is managed with the help of a search algorithm. The input of the key words by the user together with the fast, globally operating algorithm results in a capability that exceeds the standard human capability.

With the help of this categorisation, various types of artificial agents may be roughly distinguished. The categories sometimes overlap, however. For example, if a hand prosthesis, which essentially provides a restitution, is equipped with a heatproof gripper or claw, the overall capability of the man/machine system to handle hot objects exceeds that of the human hand. Hence, it may also be thought of as an expansion.

Applications of the type *substitution* are frequently monitored by a human being. The role of the human being has thus changed from that of an agent to that of a supervisor. This means that the human being is not simply replaced, but is merely given a different role. The human being, when functioning without assistance, has to supervise all activities himself; here, it is impossible for him to simultaneously provide permanent, full attention. This is one reason why the combination of a human and an artificial agent allows for a significantly higher overall capability level.

In some respects, at least concerning some of the criteria, this "substitution" type of application and the category of "expansion" overlap. However, it is still possible to use the sub-division to categorise the essential functions of the artificial agent in relation to the standard functional capabilities of the human being.

2.3
Areas of Human Replaceabilities

As shown in 2.1, artificial agents are developed in order to achieve certain aims, whereby they are, in general, but one of the means available. As it would be a comparatively complex undertaking to determine whether replacement of human capabilities should be pursued whenever possible , in an early phase, it is more expedient to tackle the subject from a different angle by posing the following question: Are there any areas in which the replacement of human being by an artificial agent is undesirable or inexpedient, and on the basis of what criteria should this be decided?[8] In a sense, this question points at various areas of non-replaceability:

---

[8] Questioning the other way round: In which areas is the replacement desirable and expedient, need not, however, lead to the same result. For assessing a certain robot in a certain context this question may be the better one. In the context of an initial assessment, to figure out general criteria of replaceability, the question stated here, seems to be most promising.

1. *Technical Non-Replaceability:* Here, it must be ascertained whether the artificial agent is able to achieve the minimum aims as defined in the catalogue of requirements. It is pointed out once more that the artificial agent does not have to achieve these aims in exactly the same manner as the human agent would. Moreover, in view of the continued development of artificial agents, the state-of-the-art is constantly changing, so that the question of technical non-replaceability will have to be reviewed again and again.
2. *Economic Non-Replaceability:* The economic consequences of replacing human beings by artificial agents should be considered from a business and an economic point of view. First, it is necessary to examine whether it is more cost-effective for a business to employ robots rather than human beings for certain tasks. In case of a break-even point and early profitability, most entrepreneurs will favour the employment of robots. This can have an effect on the entire national economy. Such macro-economic consequences must be assessed.
3. *Legal Non-Replaceability:* Legal questions arise as soon as the failure of an artificial agent raises problems of liability. Moreover, certain relations between persons such as, for example, the relation between physician and patient, or lawyer and client, are regulated by law. If artificial agents are involved in any way in these relationships, this must be held against the relevant laws (which may lead to a amendment to the law).
4. *Moral Non-Replaceability:* This refers to cases in which there is no agreement about replacing human beings for moral reasons. For example, there would be no agreement about leaving the education of children exclusively to artificial agents, even though some criteria might well be met that could not be met by human parents or teachers. For instance, artificial agents never become irritable, always have time, and may have the most up-to-date pedagogical know-how at their disposal.

3   Examples and Discussion of the Sub-divided Raster

The two-dimensional structure developed in section 2 is not specifically related to the assessment of autonomous robots in future health care. The categories can also be used to structure other fields of application. However, referring to the elements of Rational TA (Armin Grunwald in this volume), one of which is to depart from a concrete context, and to the vision introduced in the introduction, I will borrow my examples from the health care area.

An example of the RESTITUTION of human capabilities is the hand prosthesis. Artificial hands, like the four-fingered hand developed at the Institute of Robotics and Systems Dynamics of the DLR in Oberpfaffenhofen, are designed in such a way that all drives (artificial muscles) are situated in the hand or its root (Hirzinger 1996, Hirzinger et al. 1996) Micro-engineering and mechatronics have paved the way for this development. With these drives the hand can produce forces which approximately reach the order of magnitude of forces produced by natural hands. In this respect, these artificial hands may be seriously considered for use as protheses (Hirzinger 1997).

Concerning the control mechanism of the prostheses, the present state of the art allows for detection of electrical potential variations caused by muscle movement by sensors on the skin immediately above the muscle. (myo-electric control) (Bock 1995). A new approach using the still existing nerves to control the movement of the prostheses is currently under development. Firstly, an attempt is being made to connect the nerve cells to a microchip so that they may exchange signals (Fromherz and Stett 1995; Fromherz 1996; Vassanelli and Fromherz 1997). Secondly, the electrical nerve signals are detected by induction at the axons. To convert the nerve signals into commands for controlling the movement of the prostheses they are analysed by use of an artificial neuronal network. The first successes in this endeavour have already been achieved (Bogdan and Rosenstiel 1994, 1996a,b). Because of this development the technical replaceability of the human hand seems to be possible in a more efficient way in the future than at present.

Economic criteria and considerations with regard to cost-effectiveness are less relevant for consideration here. The high costs of a hand prosthesis result from the expensive hardware and the intensive rehabilitation programme. Of course, cost-effectiveness analyses are important sources of information in debates on cost control and resource allocation in health care. However, this is not the place to deal with these issues.

The legal aspects of replaceability will be confined to liability disputes arising in the event of device failure. Who is responsible for a hand failing to open, or external signals interfering with the control mechanism causing unwanted movement of the hand?

Already in the introduction a robot was discussed that is able to control the camera during minimal invasive surgery and thus provides a substitute for human capabilities. The camera is controlled by a green dot on the operating instrument of the surgeon (Arbter and Wei 1996). The position of the green dot is positioned by means of an image processing system as a marker in the centre of the picture. A tracking algorithm is able to convert the discrepancy between actual and reference marker position into command signals to the control mechanism of the camera. In this way, the green dot always appears in the centre of the picture screen. The fixed point of movement for the camera, which is determined by the opening of the incision in the skin of the patient, is regulated by the control algorithm. In this way,[9] the artificial camera assistant is technically superior to a human assistant, for there is no longer any need for communication between the surgeon and his assistant ("higher" – "over to the left"); the surgeon controls the field of vision of the camera through his operating instrument. Camera wobble caused by fatigue on the part of the human assistant is now ruled out and the fixed position of the camera is maintained automatically. Thus the robot is able to replace the human camera assistant from the technical point of view.

---

[9] According to a catalogue with technical requirements for this artefact (cp. 2.1)

From the economic perspective it should be considered whether a robot of this kind is cheaper in the long run than a human assistant doctor. After how many operations will such an investment have paid for itself? What costs are associated with training operating staff? Considerations of legal replaceability again entail questions of liability in the event of equipment failure. It might be necessary to have a second doctor in attendance at the operations. With regard to moral considerations, it should be clarified whether in general an artificial agent, acting autonomously in response to signals from a machine, should be allowed to work in an operating theatre at all.

As an example for the expansion of human capabilities we may regard the software system RIAD (or Riyadh) which was developed by René Chang (1992). It is fed from three data sets: a) a set of more than 30 relevant medical characteristics of the patient, for instance electrocardiograph readings, b) a set of details of the treatments already received from doctors and nurses, and c) a set of data to enable assessment of the illness according to an international classification system. From these data RIAD calculates the survival chances of the patient, thus providing the physician in charge with an aid to decide upon the appropriate therapy. Moreover, the programme will calculate the expected costs for the treatment of patients, give information as to whether a patient should be taken to an intensive care unit, a monitoring unit or a wake-up ward, and indicate whether or not a particular therapy will be effective. In Germany, RIAD is currently being used to assess the efficacy of a treatment by feeding the data into the system after the treatment has been completed. By means of this procedure, it has proven to be possible to evaluate the RIAD system itself.[10] The system calculated the survival chances of 53 patients (out of all patients in an intensive care unit) at 0.1%, on the basis of data entered after treatment. Of these 53 patients, 6 had in fact survived[11] (Traufetter 1997a) at a cost of more than 100,000 DM as calculated by the RIAD programme.

RIAD was conceived as a decision making tool to be used for selecting appropriate therapies. Since RIAD contains a database in which information on several hundred thousand patients are stored, it can be regarded as a technical expansion of the knowledge of the physician. However, if one considers the accuracy of the system in the example above (47 out of 53), it is clear that it is necessary to include an acceptable maximum error rate in the criteria for "technical feasibility". The economic aspects in this example are particularly interesting since in Germany, physicians is not allowed to decide between treatment options on the basis of these aspects. At the same time, from a macro-economic point of view, it is evident that medical costs have risen to their limits. Again, this touches upon ongoing debates on cost control and resource allocation in health care. For discussions on rationing, the expected costs of a given treatment would constitute a most significant factor. This relates to the legal

---

[10] By turning the argumentation the other way round. Define the achievements (survival rate) of the intensive care units as given and let RIAD calculate the survival rates afterwards.

[11] This result gave cause for vehement reactions. For example Frank U. Montgommery President of the medical association Hamburg mentioned: "If the survival of a human being is made dependent upon economic criteria, then, from the ethical point of view, we will have reached the point where the NAZIs left off."(Translation MD) (Traufetter 1997b)

questions involved. Here, it must be clarified what legal status RIAD should receive. It could be simply used as an advisory-tool, leaving the responsibility for decisions taken in the hands of the physician. However, if the rate of error could be reduced considerably, cost confinement could provide a push towards a much more important role for such a computer system. The question then remains as to who bears the responsibility for treatment decisions. From the moral point of view it will have to be clarified how the machine supported doctor's advice shall be evaluated and how the output the artificial agent should be taken into account when rationing medical services. Furthermore, the suffering of the patient is important, especially since at the end of the day, through RIAD, it may be possible to spare the patient the pain of a therapy.

The functionality of the two-dimensional raster is clear from the examples given. However, the sub-divisions should not be seen as a rigid structure. As already mentioned above, the user of an artificial hand may be in a position to handle very hot objects. Thus he would have more than standard human capabilities. Some restitutional applications may lead to an expansion of capability. Thus whatever field is referred to within the raster, the application in question will have to be defined according to its own specific criteria.

The same applies to the areas of replaceability. Here, too, the sub-divisions should not be seen as a rigid structure. As we have seen, the assessment of legal replaceability is closely connected with technical replaceability, since the capability of the artificial agent must influence the evaluation of its legal status. For example, the technical quantity "error rate" will have to be taken into account when evaluating legal aspects. Economic advantages may be confined by legal constraints, when, for example, law demands the attendance of a second doctor during minimal invasive surgical operations. Criteria of technical, economic and legal replaceability inevitably reflect moral choices. There is a link between technical and moral aspects, e.g. the relationship between doctor and patient. Also, economic aspects of rationing in health care are directly linked with moral aspects.

In fact, these connections between the several scientific disciplines are the reason for assembling an interdisciplinary expert group (section 4). Only in such interdisciplinary discussions do all participants have to uncover the underlying assumptions of their views, because they have to convince the other participants by argument. The result will be a justified argumentation chain, which is supported by all disciplines. Otherwise a Delphi procedure (BMBF 1997) would be sufficient, in which the expertise is realised by questioning all experts separately and the result of this questioning is further processed by statistical analysis. But why should the lowest common denominator reflect the "best recommendation"? (Grunwald and Langenbach 1998)

The four areas of replaceability could certainly be extended, for example, to include a social non-replaceability component in cases in which the social status of a handicapped person as part of a man/machine system is involved. The social status of a person hitherto classified as partially handicapped with special employment rights could be changed as a result of being equipped with an artificial agent. For example, by receiving an artificial hand he might become a much sought-after specialist, able to perform tasks that could not be fulfilled by others.

The aspects involved in the application of artificial agents as seen from the point of view of the user are not mentioned explicitly in this raster, but they are reflected in the criteria that define replaceability. In order to define specific replaceability criteria, it is necessary to draw on the expertise of scientists working in the different areas of application. This means that in the case of the above mentioned examples, i.e. the hand prosthesis, the camera assistant and the computer system RIAD, the criteria must be evaluated by medical experts in an expert discussion (cp. section 4) and by other stakeholders in the following steps of the TA-process.

In order to check the arranging structure of the raster it was sensible to check it by examples, which are already in use (RIAD) or in an advanced stage of development, and thus have generated useful experience and information. Artefacts or even sketches of artefacts which are parts of far future visions will also be discussible in relation to this raster. This implies that the raster could be extended or modified on the basis of arguments forwarded in the discussion. In the context of discussion, this flexibility of the structure does not invalidate the arrangement itself. Statements of the type "in the area of Restitution all developments should be recommended" are borne out by the raster, regardless of the fact that some artefacts have to be assessed on their own merits.

4 Technology Assessment

A structured, rational discussion is a principal purpose of many approaches of TA. The approach of Rational TA primarily is a rational discussion of experts as described in this section. Depending on the context, this could entail rationalising a discussion by scrutinising and structuring the arguments of the participants (e.g. climate change, cp. Armin Grunwald in this volume), or initiating a discussion in areas where TA seems to be required. Research on autonomous artificial agents has already passed the "early stages of development" (cp. Armin Grunwald 7.3 in this volume) and the societal effects that can be expected call for TA. Nevertheless, this could also be a misjudgement, i.e. the result of the discussion of the experts described below could be that no further TA is needed in this area. The expensive participation process as second step of the TA-Process could be evaded then. The framework developed so far could be the basis for discussing a vision like "In future health care most humans will be replaced by autonomous robots". Of course this is a provocative formulation, as in most views on artificial agents the notion of "replacement" has been replaced by the notion of "assistance" (cp. Peter Mambrey 2.0 in this volume). But being provocative is a permitted means to start a discussion.

The structure developed in the previous sections can be used as a starting point of a TA-project in order to assemble a group of experts in the fields discussed in section 2, namely experts of robotics, economics, jurisprudence, etc. These experts will enter into an intensive debate, which entails monthly meetings over a period of two years. The results will be recorded in a memorandum, namely a book that does justice to specific scientific requirements. Some aspects of this procedure , like the recruitment of the experts, and some rules of debate, are listed below. Further aspects of the conception of Rational Technology Assessment can be found in (Armin Grunwald in this volume, Grunwald 1998).

– The experts should be *outstanding* scholars within their scientific community. Outstanding experts can be found by monitoring the discussions and conferences of a scientific community. Who is frequently invited to panel discussions? Who gives invited talks? etc.. The reason for the requirement of being outstanding is that the expert has to represent all major paradigms in his (sub-)discipline, which is a task the experts must prepare themselves for[12]. He also has to advocate for other stakeholders connected to his discipline[13] (see below). For this purpose a kind of overall view needed.

– The experts must be interested in *interdisciplinary* discussions. This is not necessarily the same as the former point. However, very often 'outstanding' experts also feel eligible to answer questions from outside their discipline. But interdisciplinary discussion is meant in a more procedural way, which involves, for example, signing a consent form that states rules of debate. An important point is that all statements must be supported with comprehensible evidence and arguments. In an interdisciplinary debate this means that the arguments must be comprehensible to all participants. Another point is that concrete recommendations for action should be given. Dissenting votes should be exceptional and have to be supported by strong arguments as well. Due to the fact that, in general, every scientist is an expert in just one scientific field, these rules presuppose a very intensive debate. Not every expert is prepared to participate in such a procedural debate.

– The selection procedure generally leads to several experts, of whom one or two per discipline or sub-discipline will be appointed to participate in the project group. Having equal and small numbers of experts out of every scientific community is not only sensible for organisational reasons, but also helps to avoid dominance of some disciplinary views. Every expert has to convince the experts of other disciplines of his argumentation. So, in the ideal case, the resulting memorandum will not be biased by the aims of one specific discipline or particular interests.

– The previous point results in the participation in the project of just one or two experts per discipline and naturally these experts influence the project in a personal way, even when they are asked to represent all major trends and paradigms in their discipline. The points of view of these individual experts can be better appreciated by organising additional meetings (KickOff- or MidTerm-Meetings) where *other* experts get the possibility to place their opinions on

---

[12] However, there is always some doubt left, as to whether experts will defend paradigms of others with the same accuracy as their own ones. Therefore other experts are invited to forward their arguments at supplementary meetings (Kick-off, Midterm, etc.).

[13] Here, I do not claim this advocacy to be complete (Cp. Rob Reuzel and Gert Jan van der Wilt's Chapter). See below for the consideration of stakeholders.

record. The arguments of the external experts have to be taken into consideration by the expert group in each case. Either by including the argumentation or by justified non-consideration.

Autonomous robots represent the main topic in a vision "In future health care, most humans will be replaced by autonomous robots" and generally the discussion about the artefacts is a good starting point for the discussion about a vision underlying these artefacts. Nevertheless an assessment of a vision should not be reduced to the assessment of the artefacts contained therein. The next step in the TA-process regarding robots in health care will be to study the consequences for future society. What secondary skills are shown by human beings, when they fulfil the tasks that should be taken over by autonomous robots? How should these additional skills be valued? In the case of the camera assistant, these skills are well defined. One opinion could be that being "assistant physician", which, for example, includes advising the surgeon on decisions during operation, is the primary skill and "handling the camera" is the secondary. This would imply that both the assistant physician and the autonomous robot should assist the surgeon. In the case of a service robot helping the nursing staff with serving and feeding meals the skills are less well defined. How should some additional words of comfort be valued and how do they support the healing process?

Changing the point of view one could question the benefit using artificial agents. In the case of the camera assistant the benefit is clear to the surgeon, who is able to obtain better results, and the patient, who is also interested in better results. In the case of the autonomous robot feeding patients and serving meals, the benefit seems to be absent at first sight, assuming that patients prefer human service. What are the requirements and the demands on artificial agents made by handicapped people or in-patients and what are the reasons for these demands? Due to the fact that the assistance of an artificial agent is variable and adjustable, the level of assistance can be chosen by the users themselves to meet their wishes. For the nursing staff, such individual assistance adapted to the abilities of the patient is, in general, not feasible due to lack of time. The advantage for the patient would be more autonomy, perhaps resulting in a higher level of self-esteem. How could this be rated and how would it support the healing process?

In the context of these questions, the demand on the experts to be a kind of advocate for other stakeholders can be made more explicit. Within the interdisciplinary expert group, the medical expert has to deal with both the evaluation of technical replaceability, by setting up the catalogue of requirements and assessing the robot against this catalogue, and the evaluation of patient benefit. So the stakeholder "patient" is present only in an indirect way. I propose the direct participation of non-expert stakeholders takes place after[14] the expert discussion. For example, the British Deaf Association, in the case of cochlear implants discussed by Rob Reuzel and Gert Jan van der Wilt, would be invited to participate in a second phase of the TA-process (see below). The medical expert

---

[14] "After it" is meant only in a temporal regard, as for example the second step in the TA-Process, and not in a hierarchical sense.

tries to anticipate the point of view of these stakeholders in his contribution to the memorandum.

The interdisciplinary debate on the artefacts and their consequences will transform the vision the process departed from in a more balanced one due to following reasons (Gethmann 1979; Gutmann and Hanekamp 1998; Rescher 1988):

- Every expert is an expert just in his discipline. Therefore, an interdisciplinary debate starts with discussion about notions, their definitions and underlying assumptions, that are common in particular disciplines and therefore are rarely discussed. In this way, underlying assumptions are made explicit, because they have to be explained to experts of the other disciplines.
- After this process synthesising the views of the different disciplines is done by criticising the other views. In this phase, the development of transparent argumentation chains starts, where the arguments come from the different disciplines and have to be accepted by the other disciplines. This results in one or more chains of argument supported by all participants.
- In the last phase, these argumentation chains have to be extended into the area of normative aspects, i.e., the initial impulse of the research project. Again this should be done by reference to rational arguments. In this constructive phase, the experts have to decide on recommendations for acting. In the procedure described so far, this involves selecting one argumentation chain, consisting of a descriptive and a normative part, out of all those developed so far.

Summarising the above reasoning and translating it into a procedure leads to the following steps. An initial disciplinary vision is taken as a starting point. Other disciplines are asked to articulate their views on this vision. In an interdisciplinary debate an unbroken chain of argument is developed and presented in a transparent manner. This chain includes a recommendation for to acting[15]. In this sense, the memorandum is a kind of vision labelled "optimised for understanding and, therefore, open for criticism from other parties".

These other parties should include both experts with views not yet represented, as well as interest groups and individuals who may have a stake or are otherwise involved in the technology (see below). Criticising the memorandum is the first step in the second phase of the TA-process. Later on, this criticism has to be transformed in a constructive process, in which again one argumentation chain has to be selected, consisting of a descriptive and a normative part. It might be true that several possibilities remain, each bound to particular conditions. Also, it is possible that solutions are amended in the process, and even that new ones emerge: the deliberation between the original experts and new participants may improve the quality and scope of the solutions as well as yield new creativity.

In the case of the cochlear implants discussed in the contribution of Rob Reuzel and Gert Jan van der Wilt in this volume, there are well organised target groups,

---

[15] These recommendations to act given by the group of experts have also the aforementioned shortages like "dependent on the experts" and "experts stand in for other stakeholders". But nevertheless the transparent transformation of arguments into recommendations is important and will be criticised in the following steps as well.

for instance, the deaf organisations. In other areas, most probably in the case of robots in operation theatres, there are no organised target groups, perhaps due to the fact that advanced technology is already common in operation theatres. But health care organisations, organisations of handicapped (which are directly involved in some cases), medical associations and health insurers will formulate their opinions, views, and visions. Visions from other contexts may already exist (and must be taken into consideration by the expert group). There may be overlap with other subjects like tele-medicine or decentralised virtual health care. A new visionary memorandum will be considered and commented on by these scientific networks anyway.

In addition, the participation of laypersons could also be a well-tried procedure to obtain useful criticism. In some contexts, where individual laypersons involved can be found (compare the example of the serving and feeding robot above), this would complete the memorandum in a special sense. In the case of the vision "Autonomous Robots in future Health Care" so far no in-depth discussion is taking place and thus, in general, these "organisations of concerned" do not exist. Nevertheless, nearly everybody has heard about autonomous robots in science fiction literature and movies. Therefore, it seems to be possible to evoke reactions from individual stakeholders to the scientists' vision. This provides an occasion for stakeholders for scrutinising the way they were represented by the experts and bringing in additional arguments. The memorandum of the expert group could be the basis of discussion with laypersons as well, because, in general, the argumentation is more transparent than in disciplinary debates. After all, as already said, no scientist is an expert for all disciplines. It would be of advantage to include[16] the same experts in the debate with laypersons for three reasons:

1. Some arguments forwarded by the laypersons will already have been discussed in the interdisciplinary expert group. Thus the argumentation and the discursive channels leading to this argument will still be present.
2. Some aspects forwarded by the laypersons are based on emotions (cp. Armin Grunwald 5.1 in this volume) and one has to question whether there are good or bad arguments behind these emotions. Due the intensive debate within the expert group, the experts will, however, be in a position to formulate the rational arguments behind an intuitive statement of laypersons. Here, it is not meant that the experts should spoon-feed the laypersons. Due to the fact that an expert is only an expert in his own scientific field, in an interdisciplinary debate also emotive statements are made. The expert[17] such a statement is addressed to has to figure out which rational arguments could be behind such a statement. It is necessary to determine whether the arguments in the debate were already considered or not.

---

[16] The discussion scenario described here involves the experts directly in the discussion of laypersons. There is also the possibility to start an iterative process in which the laypersons work out their point of view and the experts take this into account in their next round of discussion and the other way round.

[17] The expert for the field to which the statement is related.

3. Some arguments forwarded by laypersons perhaps have not been considered in the debate so far[18]. Due to the fact that they still have their discussions in mind, the group of experts may be able to judge ad hoc on these new arguments from their point of view. However, if they do not have strong arguments for refusing them, they cannot do so.

The results of the participation process on the basis of the expert memorandum could complete the vision "Autonomous Robots in future Health Care". The findings of this process could be useful in other areas as well, and used to develop to the core concept of, say, a vision like "Autonomous robots in future society". Within this vision the co-operation of humans and autonomous artificial agents in general will be in the focus of interest. The different parties at the meeting will have to prepare themselves for optimising co-operation. On the part of the developers of robots that could mean that, for example, models of non-expert users be designed and that the artificial agents be equipped for intuitive operation by humans. On the part of the humans, this could mean that a learning process must take place. In addition to these direct consequences of the contact with autonomous robots, one has to outline what the indirect consequences could look like. Why is the service area a major field of application of artificial agents? What are the users looking for? An infinitely variable, adjustable level of assistance? Incessant friendliness and reliability? This kind of service obviously cannot be implemented by free humans, who are not adjustable and who are subject to changes in mood. Which societal modifications would be the result of such a non-human service? What happens to societal communication when all services so far produced by humans will be made available through autonomous robots? Perhaps the responses to these questions will initiate a learning process as well. In any case, they should be scrutinised by Rational Technology Assessment (Armin Grunwald 5.1 (Rationality) and 5.2 ("understandable for everyone") in this volume) and the project proposed in this contribution could be a first step.

5   Conclusion

In this paper, I suggest that a team of experts in many cases provides a good starting point for TA. On the one hand, there are areas, like climate change or nuclear power, in which already very much is discussed and where several schools of thought have been formed. On the other hand, a new topic, and the replacement of human beings by autonomous robots is an example in this respect, should be tackled by TA. In both cases the first step should be to initiate a rational interdisciplinary debate to structure the arguments in the ongoing discussion or to structure the new field of TA. The strict rules of the debate should ensure transparent representation of the results in the sense that the results are optimised for understanding and, therefore, open for criticism from other parties. The

---

[18] For example in that cases, where the expert is acting as an advocate for other stakeholders (cp. Rob Reuzel and Gert Jan van der Wilt in this volume).

detailed description of the reasoning process makes the resulting memorandum valuable in further steps of TA, whether it is used for further discussion within the science community, within policy, or within a participation process.

References

Arbter K, Wei GQ (1996) "Verfahren zur Nachführung eines Stereo-Laparoskope in der minimal invasiven Chirurgie" German Patent, No.3943917, applied Aug.14, 1995,issued July, 1996; U.S. Patent, Pending; French Patent, Pending

BMBF (1997) Delphi '98 – Studie zur globalen Entwicklung von Wissenschaft und Technik. Bonn

Bock O (1995) Entwicklung, Fertigung und Praxis in der Armprothetik, Produktinformation, p 8

Bogdan M, Rosenstiel W (1994) Artificial Neural Nets for Peripheral Nervous System - remoted Limb Protheses, Proceedings of NeuroNimes'94, Marseille, France, p 193

Bogdan M, Rosenstiel W (1996a) Classification of Nerve Signals using Kohonen`s Self-Organizing Map, *First International Conference on Bioelectromagnetism*, Tampere, Finland, p 239

Bogdan M, Rosenstiel W (1996b) Intelligent Neural Interface - Signalprocessing of Nerve Signals using Artificial Neural Nets, BioNet'96 (Invited paper), Berlin, p 33

Chang R (1992) Riyadh ICU Programm ™, Produkt Information

Decker M (1997) Perspektiven der Robotik. Überlegungen zur Ersetzbarkeit des Menschen. Graue Reihe Nr. 8, ISSN 1435-487X

Dewey J (1922) Human Nature and Conduct. New York

Fromherz P, Stett A (1995) Silicon-Neuron Junction: Capacitive Stimulation of an Individual Neuron on a Silicon Chip. In: Physical Review Letters, Volume 75 No. 8:1670-1673

Fromherz P (1996) Interfacing Neurons and Silicon by Electrical Induction. Ber. Bunsenges. Phys. Chem. 100, pp 1093-1102

Gethmann CF (1979) Proto-Logik. Untersuchungen zur formalen Pragmatik von Begründungsdiskursen. Suhrkamp, Frankfurt

Grunwald A (1998) (ed) Rationale Technikfolgenbeurteilung, Konzeption und methodische Grundlagen. Springer, Berlin Heidelberg New York

Grunwald A, Langenbach C (1998) Die Prognose von Technikfolgen. Methodische Grundlagen und Verfahren. In: Grunwald A (ed) Rationale Technikfolgenbeurteilung. Konzeption und methodische Grundlagen, Heidelberg, Springer, pp 132-156

Gutmann M, Hanekamp G (1998) Wissenschaftstheoretische Grundlagen Rationaler Technikfolgenbeurteilung. In: Grunwald A (ed) Rationale Technikfolgenbeurteilung. Konzeption und methodische Grundlagen. Springer, Berlin Heidelberg New York, pp 55-91

Hirzinger G (1996) Mechatronik, 3D-Graphik und Telepräsenz – Neue Anstöße für Maschinenbau, Robotik und Medizintechnik. VDI Berichte Nr. 1270:26

Hirzinger G, Koeppe R, Baader A, Lange F, Ralli E, Albu-Schäfer A, Staudte R, Wie W-Q (1996) Neural Perception and Manipulation in Robotics, Statusseminar des BMBF: Neuroinformatik und Künstliche Intelligenz

Hirzinger G (1997) Roboter mit Orientierungssinn, Frankfurter Allgemeine Zeitung Nr. 30 , p N1

Janich P (1993) Mensch und Automat. In: DLR (ed) Bemannte Raumfahrt im Widerstreit (Kolloquium am 14.4.1993 in Bonn). pp 25-34

Paschen H, Petermann Th (1991) Technikfolgenabschätzung - ein strategisches Rahmenkonzept für die Analyse und Bewertung von Technikfolgen. In: Petermann Th (ed.) (1991) Technikfolgen-Abschätzung als Technikforschung und Politikberatung. Campus, Frankfurt, pp 19-42

Rescher N (1988) Rationality. Cambridge

Rohbeck J (1993) Technologische Urteilskraft. Zu einer Ethik technischen Handelns. Frankfurt

Traufetter G (1997a) Der digitale Todesbote, Die Woche 28.2.1997, p 26

Traufetter G (1997b) Kalkulierter Tod, Die Woche 7.3.1997, p 30

Vassanelli S, Fromherz P (1997) Neurons from rat brain coupled to transistors, Appl. Phys. A 65, pp 85-88

IV   Conclusions

The Lessons we Learnt:
First Outline of Strategy and a Methodical Repertoire for Vision Assessment

John Grin, Armin Grunwald, Michael Decker, Peter Mambrey, Rob Reuzel, Gert Jan van der Wilt

1   Introduction:
Modern Visions and Societal Structures

As the introductory paper by John Grin put it, this book is about ways to think, as well as on ways to think about thinking, about the future, focusing on the relation between societal problems and technology. In particular, we have focused on the role technology assessment (TA) may play in assessing the visions that are guiding the ways in which actors, in specific sectors, shape their segment of 21st century society through their collective actions. Specifically, the undertaking reported here was inspired by the suspicion of at least one of us that many so-called revolutionary visions for the 21st century are, on the level of their basic assumptions, not that different from the visions that have dominated over most of the 20th century. That is, they too reflect those assumptions that are so typical for High Modernity (see the table in section 4 of that paper). At the core are the assumptions that social progress can be obtained through sound and certain, scientific, knowledge and its application in technology; and that, therefore, society should be guided by institutions that are able to translate such knowledge into courses for action.

What we have seen in investigating some current visions in the preceding chapters is an intriguing and complex mix of, on the one hand, a dominance of elements that represent High Modernity's assumptions with, on the other hand, elements that reflect other biases. Put somewhat provocatively, the nuanced view preferred by Professor Louise Fresco concerning the world food problem (chapter 1) may indeed be seen as a 'point representation' of society's mix of biases concerning technological visions. In this mix, dominant modern elements of visions are complemented by differently biased elements, thus correcting some of the blind spots and risks of modern views. As should not surprise us, we have seen indications that this reflects the way in which socio-technical systems, having emerged since Enlightenment and, especially, since the Industrial Revolution, tend to reproduce themselves.

For instance, in the chapter by Rob Reuzel and Gert Jan van der Wilt, we have encountered Cochlear Implants. As a solution for deafness it is both being envisaged by *and* presupposing medical-technical expertise. That is, this development is a typical reflection of the emergence of a medicalised and professionalised health care system. Rob Reuzel and Gert Jan van der Wilt cite work by Barbara Duden and by Michel Foucault who have documented the gradual development towards ensuring health through a professionalised health care system.

Yet, in addition to this solution also critics can be heard. Most interestingly, much of the criticism, brought forward e.g. by deaf people's organisations, does not focus on the artefact *per se,* but rather on the 'anti-vision' that critics suspect to be embedded in it: the idea of a 'deafless society.' To be sure, proponents would not consider this as an appropriate designation of their vision. Rather, they would say that they are driven by an attractive, healing vision in which deafness as a handicap is eradicated. Deaf people's organisations tend to be more critical and see it as a way of excluding deaf people from 'normal' society. The reasoning behind this fear is that within the current health care system, with its emphasis on expert-based central distribution of scarce resources, funding such a relatively expensive treatment may drive out alternatives. Thus, this criticism reflects the fear that within existing structures, action will be pushed into a direction fitting the dominant experts' views, but much less deaf people's viewpoints. Feeling uneasy about a too one-sided reliance on technology, they plea for a solution in which a varied repertoire of measures (sign language, cochlear implants and so on) remain available to curers, carers and deaf people. Their agenda is to put cochlear implants in some broader framework so as to prevent this anti-vision from arising. Or, formulated more positively, they may accept cochlear implants, but only if it is embedded it into a vision that they may appreciate better than that of a 'deafless society'.

It is of interest to emphasise that their concerns are rooted in the development they expect, given the structural properties of the health care system, to be triggered by the further development of this technology. It must be admitted that it would be difficult to hold that existing decision making routines bear many safeguards against such risks. To the extent that these procedures (implicitly) exclude rather than deliberately take into account other than medical-technical expertise, there is indeed reason for doubt on their capability to define a more balanced vision. Elsewhere, Gert Jan van der Wilt (1995) has demonstrated how this may lead to rather one-sided orientations of medical technologies. He shows how by far most resources spent on Parkinson's disease are devoted to one particular combination of treatment and diagnosis which, moreover, focuses more on restoring assumed dysfunctioning in the brain than on solving patients' daily bothers.

Viewed in these terms, John Grin's chapter on the vision of a 'bloodless war,' deals with matters that are similar in several respects. First, weapons for 'mass protection' are not rejected as such. Few people will dispute the use of technologies that may achieve casualty reduction. Here too, criticism is against the vision in which they are – not only allegedly, in this case – embedded: that of a bloodless war as a response to the problem of maintaining public legitimacy of humanitarian and other interventions. Integrating technologies that may promote casualty reduction into a vision that includes a variety of elements, rather than merely technical fixes would meet such criticism.

On the other hand, this chapter also contains an example of a technology that is criticised *per se*: Joint Stars, a system to enable deep attacks on adversary assets so as to prevent them from engaging in deadly ground battle. Here, the criticism is against the typically modern emphasis on technological progress as a universal solution, without taking into account contextual circumstances. But even here, the standard against which the impact of context on system effectiveness is judged derives from the vision to which the system is supposed to contribute.

How such typically Modern ideas have historically developed within the world of airpower has been discussed too in his chapter. It is a process in which expectations concerning maximisation of speed, lethality, acceleration and range have governed both technological and conceptual development ever since the emergence of airpower. Over the decades, a succession of visions has arisen from these expectations. May be the most dramatic illustration how strong this bias is concerns the way in which defence ministries tend to respond to the cost increases associated with the 'baroque arsenal' phenomenon. Apparently neglecting the roots of the cost dynamics, international co-operation is sought as a solution, which more often than not makes the arsenal even more 'baroque' (with rococo tendencies, one might add) and costly.

The second major similarity with the cochlear implant case is that, also in this case, one mode of expertise seems to dominate decision-making routines. Studies of existing procedures, as well as of attempts to broaden them, indeed confirm this (van Houwelingen 1992; Hatchett and Reuter 1992; Enserink 1993). Attempts to broaden decision-making, so as to include a wider variety of viewpoints and aspects (such as stability and arms control), tend to encounter either of two fates (Brauch *et al.* 1997):

- *Either* the dominance of one type of military thinking within defence policy making implies the rationale or necessity to bring in other viewpoints from institutional loci outside the regular decision making process. The arms control impact statements done between 1977 and 1994 in the United States by the State Department are a case in point. The reverse side of the coin is that these statements appeared to have little effect on the defence acquisition policy: Congress usually paid little attention to them when taking decisions on materiel acquisition.
- *Or* aspects like arms control and stability are considered within the regular decision making process, but are being operationalised and judged from the perspective of dominant military thinking (and not always as sympathetic as Louise Fresco considered different visions in her lecture). This happened with the sections on arms control and stability in the documents produced in the so-called defence materiel choice process in the Netherlands, designed to increase political control over the acquisition process, particularly including security political considerations other than military effectiveness.

Peter Mambrey's chapter on assistant computers illustrates other mechanisms than central decision making routines that give rise to tendencies of socio-technical systems to reproduce themselves. Here too, as we have seen, the 'flavour' of the technological routes taken reflects the dominance of High Modernity assumptions. Peter Mambrey shows that the vision embedded in work on the 'assistant computer' has hardly been influenced by more recent discourse on organisations, man-machine interfaces and so on. Information scientists appear to be guided by implicit views by more classical visions. This becomes may be most clear from the rather technology centred (as opposed to human-centred) operationalisation of the notion of computer assistant. More subtly, the assistance tasks were focused on well-structured tasks rather than on problem structuring. Thus co-ordination was reduced to effective task fulfilment of each organisational member, and less on supporting the tuning of complex organisational action, taking into account its organisational context, including the power relations therein.

The implicit dominance of High Modernity's assumptions appears to be rooted in the fact that the designers of computing systems appear isolated from recent discourses in social science, making them hold widespread, more modern assumptions. As the bibliographic notes in Gareth Morgan's (1997, pp 379-431) already classical discussion of different metaphors for organisations suggest, more recent metaphors correspond to the type of criticism of High Modernity discussed in chapter 1 by John Grin. Thus, in the case discussed by Peter Mambrey, the mechanism through which existing visions reproduce modern assumption are rooted in the structural separation of technology development from a more plural debate on the assumptions underlying it. While the process occurring here is more decentral and diffuse than those in health care and defence planning, the similarity is that technological decision making structurally privileges expertise that follows classical assumptions, while other viewpoints have considerably less access.

A full appreciation of this phenomenon implies a fuller appreciation of the scope of the challenge formulated in the introductory chapter: attempting to identify ways and means to reshape visions so as to become more balanced in terms of their underlying assumptions. To realise these objectives, it was noted, political judgement is necessary. But the above understanding of the mechanisms through which visions tend to reproduce High Modernity's basic assumptions focuses our attention to the nature of such political judgement as a deliberate attempt to re-orient existing visions in a more or less fundamental sense.

To explore the challenge a bit further in these terms, it is useful to consider it in terms of Anthony Giddens' (1984) structuration theory. Central to that theory is the notion of the 'duality of structure': "the structural properties of social systems are both medium and outcome of the processes they recursively organise" (1984, p 25). The structures that we experience today have been formed in the past and are being influenced by our current action. Structures are not 'external' to individuals; to the extent that they shape action this is through the way in which intentionally acting individuals interpret them as enabling and constraining. Yet, human knowledgeability is always bounded, and

> the flow of action continually produces consequences which are unintended by actors (…) [H]uman history is created by intentional activities but is not an intended project; it persistently eludes efforts to bring it under conscious direction. However,

> such attempts are continuously being made by human beings, who operate under
> the threat and the promise of the circumstance that they may be the only creatures
> who make their 'history' in cognisance of that fact. (ibidem, p 27)

The duality of structure, thus, implies the recognition that structures, through their
effects on action, tend to reproduce themselves; the ways in which Modern assumptions are being reproduced in the visions discussed above may be seen as
illustrations. Dialectically related to that recognition is a second recognition: that
structures can only reproduce themselves through knowledgeable human beings.
As Giddens (1984, pp 27-28) puts it somewhat later:

> [But] in many contexts of human life there occur processes of selective 'informa-
> tion filtering' whereby strategically placed actors seek reflexively to regulate the
> overall conditions of system reproduction either to keep things as they are or to
> change them.

May be Fox and Miller's (1996, p 91) formulation of this corollary of the duality
of structure can make the implications for this book's project clearer.

> The inevitable evolution of recursive practices usually happens as the result of un-
> intended consequence and the permeability of given clusters of recursive practices
> to changes initiated elsewhere. But these may also be adjusted by discursive will
> formation.

In these terms we are, ultimately, looking for ways and means for TA to support
such critical will formation through introducing a 'critical discursive moment ' in
a recursive practice in order to adapt actions within this practice in terms of the
fundamental assumptions underlying them. That is, we are looking for TA to become a tool for bringing a wider variety of voices into the places where socio-
technological visions are being shaped and translated into technology development.

In the next section, we will explore this issue further by looking at it from the
viewpoint of the tension between the need for and the practical and normative
intricacies of long term planning. May be the most interesting insight that the
preceding discussion contributes to that exercise is the realisation that recursive
practices are both a necessary and a natural element in the constitution of societies, so that long term guidance can but be grounded in such recursive practices.

In order for TA to contribute to re-constructing visions as a 'critical discursive
moment' in socio-technical practices, we need to deal with several methodical
questions:

- How can vision assessment contribute to long term planning?
- How to uncover assumptions underlying visions?
- How to critically assess visions in terms of a variety of criteria, reflecting different basic assumptions; and how to proceed, from such critical assessment, towards a type of constructive assessment that may help to shape the future?

In the final three sections of this chapter we will attempt to answer these questions
– or, more accurately, our discussion is intended to indicate the contours of a rep-

ertoire of answers from which the reader may choose in dealing with a particular vision. We will try and formulate this repertoire through reviewing what the preceding chapters have taught us.

2   Vision assessment – a Contribution to Enable Long-Term Orientation of Technology Development and Policies

As was noted in John Grin's introduction, long-term considerations concerning ideas of the future society we wish to live in, of technological systems which seem to be appreciable to have in the future, of the development of the self-image of human beings, of possible directions of the "co-evolution" of technology and society (Constant 1978) share one particular fate. They are often suspected to be "utopian" ideas, normative of nature and showing inherent tendencies towards authoritarian thinking. This suspicion, while not new, has doubtless be reinforced by the *Zeitgeist* of the late 20th century. That spirit has been co-determined by the deterioration of orthodox marxist ideals – due more to the recognition of the internals of such regimes in terms violation of basic rights than by its socio-economic failures - as well as by the erosion of the western "ideal" of an unlimited growth of economy and individual welfare. These developments have made many people to assume that one could but renounce long-term-considerations or normative ideas on the human and societal future *altogether*. In this spirit, there is unmistakably a revival of incrementalist approaches to dealing with future affairs. Thus there is, *de facto*, renewed interest in following, more or less closely, Popper's ideas of a "muddling through", handled by the "piecemeal engineer" (Popper 1969), which was ingeniously designated by Lindblom and Braybrooke (1963) as "disjointed incrementalism" (discussed in the chapter by Armin Grunwald, sections 4 and 7).

On the other hand, it was noted in the introductory paper by John Grin that merely rejecting utopian ideals for their authoritarian features misses the drama: the fact that not all of us are unambiguously happy with the world as it is or as it is envisaged to be. The preceding chapters have made clear how, in a variety of sectors, stakeholders show an interest in shaping the ways in which technology development relates to social problems – or in the terminology of the end of the previous section: in realising 'critical discursive moments' in existing practices. Acknowledging this widely felt need, scholars in technology studies have over the last decade emphasised the shaping of technology (e.g. Bijker and Law 1994; Rip et al. 1995). This contrasts the idea of merely adapting society to the requirements of the technologically induced change (as a "technology determinism" approach would imply, cf. Grunwald 1998). This shift, together with the recognition that the ambition to predict is bound to fly in the face of the fundamental constraints implied by limited information and bounded rationality (Grunwald and Langenbach 1998), has led to a shift in focus in the field of technology assessment (Hoppe and Grin 1995; Grin and van de Graaf 1996).

Partly against this background, it is now rather generally being agreed upon that TA has explicitly to pay attention to normative and political aspects (Van Eijndhoven 1995; 1997; Paschen 1999). Challenges in shaping technology touch upon questions such as what kind of society we want to live in, what self-image of man we are thereby inferring and whether, or under what conditions, this, in turn, is desirable (Grunwald 1996). Shaping technology, therefore, often leads to morally relevant political challenges and as such involves the necessity of explicitly taking into account normative reflection in TA processes (Grunwald 1999).

In his chapter, Armin Grunwald has undertaken the quest for a more solid and normatively sound basis for shaping, in non-authoritarian ways, our future. He has stipulated some additional, fundamental considerations why the desire to shape technology implies that one cannot renounce long-term considerations, including normative aspects:

1. The functional argument: many policy problems imply a requirement for a long-term perspective. Examples are well known even in traditional technology policy, especially in cases where long lead time R&D is involved (up to 20 to 25 years in the field of materials research, Harig and Langenbach 1999). They include the challenge to avoid blind alley developments and the necessity to ensure planning security for investors over a certain amount of time as a precondition for investing. Also, as discussed in Grunwald's chapter, realising the agenda of sustainable development implies the need to adopt some long-term orientation.

2. The culturalistic argument: the inherent rationality of any culture leads to a non-arbitrariness of acting and deciding (Schwarz and Thompson 1990; Hartmann and Janich 1996). Shaping the future depends on assessing the past, the continuity and stability are pre-conditions for establishing a culture at all. Thus, reflections on the future are an important element to ensure the coherence of the expectations on the future, the needs of the present and the experiences from the past, which seems to form an essential element of cultural identity.

As Armin Grunwald emphasises, these arguments imply that a purely incrementalist approach is not sufficient. We need at least add some normative guidance to incrementalist prescriptions, based, amongst other things, on long-term considerations and aspirations. A purely incrementalist approach to the future would imply the abandoning the ambition of shaping at all and would destroy the coherence of the past, the present and the expected future: we would be losing the future instead of creating it.

In this book, we have focused on one concept that may provide normative, long-term orientation: the visions (Leitbilder) guiding, amidst other factors, technology development (Dierkes et al. 1992; Mambrey et al. 1995; Peter Mambrey in this volume). Visions – as certain forms of long-term considerations – are important elements of stabilising future expectations because they are

1. trans-subjective in the sense that visions are shared, as a rule, among the participants in ongoing developments, among particular groups of society or even in society as a whole, and
2. being followed - as guiding visions - in concrete practices of technology development (Weyer et al. 1997; Dierkes et al. 1992; Mambrey et al. 1995; Peter Mambrey in this volume).

In this way, they achieve their capability in shaping the future. Visions are relatively stable. As we have seen in the previous section, they are rooted in our culture, our traditions and morals – and thus, they are related to our past (cf. the introductory chapter by John Grin and section 6 of the chapter of Armin Grunwald). They both guide and are being maintained in recursive practices. In the sense of the directed incrementalism: the cultural coherence is materialised in visions; visions relate the past to the future.

In these respects, it may be time to rethink the risks of authoritarian visions. Of course, the collapse of orthodox Marxism teaches us the risks of authoritarian visions being forced upon societies, and the failure of the capitalist notion of progress shows how too one-sided visions may blind us for particular risks. But the challenge we are facing in this book is not so much to protect ourselves against the emergence of new, authoritarian or one-sided visions. Existing strong and continuously reproduced visions simply cannot be so rigorously replaced. The challenge rather is to transform these visions into ones that are favoured by a broader set of stakeholders, and that bear less risks for the future. So if we are dealing with the risky power of visions, it is first and foremost that of existing visions, deeply embedded in our culture.

Seen this way, it is quintessential to critically judge and re-shape visions, balancing pluriformity and the need for binding decisions, contextualisation and decontextualisation, continuity of culture and the flexibility of society - all to reach cultural coherence, as a regulative idea. Visions, in this context, are materialised results of such reasoning processes, an important medium of society coping with shaping the future and, simultaneously, taking into account the present and the past. In these processes, different actors may each have their own vision. Even if these by themselves, unilaterally implemented, would have authoritarian consequences, the process should ensure that the end result is a 'fusion' of visions into one which takes into account a wider set of normative considerations.

In this light, incrementalism may be re-appreciated and re-defined. If properly conducted, it both offers a pragmatic way out of the rule of existing visions and protects us against the reign of any singular vision. The concept of directed incrementalism, developed by Armin Grunwald, precisely aims for that. It offers a strategy in which TA as vision assessment, discussed by Peter Mambrey and others in this book, may be embedded. Thus, before going into a repertoire of methodical advices for TA as vision assessment, we offer an outline of that strategy.

How to give Normative Direction to 'Directed Incrementalism'?

At the heart of directed incrementalism is the idea that the basis for assessing visions can be found in the "cultural rationality" which may deliver a sufficient amount of stability and continuity (as is shown in the chapter of Armin Grunwald). This rationality yields a pre-condition of recognising changes as such and of relating changes to their context. It helps to identify in what aspects and relative to what criteria changes are really perceived as changes, and where the stability conditions are which allow to detect such changes – cultural rationality as the *tertium comparationis* of talking about social changes.

Cultural rationality is not changing as rapidly as is often claimed referring to technologically induced accelerations of social change. Instead, there are incremental processes to become familiar with new technology. Mostly, technology changes the "surface" of societies but not instantaneously its inherent rationality standards. Therefore, and in that sense, as we have seen, the suspicion that the visions for the 21st century are not so much differing from those which have governed the 20th century has proven correct. The technologies as the "materialisations" of the visions proposed have changed. But beyond that the old ideas of justice, freedom, division of power, freeing humans from hard work, enabling communication as the results of the Enlightenment are still working: old ideas in new frameworks. It is not surprising that going deeper into such questions allows to grasp longer time scales: at the surface there is a rapid varying process but at other levels of society processes are running more smoothly and require much more time. Exactly this situation allows for vision assessment at all.

From the review of the preceding chapters in section 1 of this paper we also see that this stability concerns, especially, the dominance of High Modernity's assumptions. When it comes to the ways in which these precisely manifest themselves, it appears that this differs among sectors.[1] As a consequence, opportunities and constraints for re-shaping visions may also differ between sectors. Another corollary is that also the substance of cultural rationality may, at least partly, have to be understood contextually. For instance, the notion of human dignity, laid down in the German Constitution, may be operationalised for decision making on medical R&D through notions that are specific for the health care system. These notions may be codified in e.g. assessment criteria in official medical TA criteria or in the legislation governing benefit packages. On the other hand, for matters concerning ethnic cleaning or other types of severe minority rights; violations, the operationalisation of human dignity may be found in publicly shared principles of protection against life threats and discrimination, encoded in the UN Charter, the Genocide Treaty and other pieces of international law.

But whether on the level of society as a whole, or on the level of a sector, and whether in one particular way or in a variety of ways (cf. section 4.2 of the introductory paper by John Grin), we do need to define *into what directions* precisely adaptation of existing visions is sought. Several answers can be found in TA literature:

---

[1] In the next section, we will investigate where to look for these manifestations.

- Normative philosophical principles, like the categorical imperative or Bentham's rule, or those implied by theories of justice;
- Value research, on the basis of insights from by the social sciences, could investigate the moral preferences and the values accepted and acknowledged in society, and use them, e.g., to see whether or not pluriformity is being sufficiently warranted;
- Citizens could produce normativity through participating in certain public discourses;
- Politicians, as democratically legitimised representatives of the people, decide.

What answer is most appropriate is the complicated issue of reaching acceptable closure in normative disputes. Ultimately, it seems to us, it is a matter of political choice (cf. van de Graaf and Grin 1999). Even the first two answers, seemingly analytical as they are, require some decision on how exactly to apply them: generally, they require some choice on how to operationalise (that is, interpretatively apply) principles, methods or prescriptions in particular cases. Similarly, the results of citizen participation depend both on the conditions under which the deliberation takes place, and on who is participating. Especially the second factor cannot be unambiguously defined, however, and thus presupposes some choice.

An Outline of the 'Directed Incrementalist' Approach

The following points may serve to summarise the approach of directed incrementalism, enriched with illustrations from the other chapters:

1. A pragmatic understanding of the "long" in long-term considerations is required. What is meant by talking about the "long run"? How long is long? What is the lifetime of a vision? The answer depends on the cultural state and the dynamics of society: it is itself a parameter to be contextually defined. For example, the visions guiding the construction of the grand cathedrals in the medieval age had a considerably larger lifetime than the vision of the "paperless office". In the present situation, at the threshold of the 21st century, the following characteristics may be valid for many cases: visions relevant for technology policy-making should claim to hold longer than short-ranged moods and should have a lifetime at least comparable to the time frames of R&D processes.

2. This pragmatic understanding of "long" leads to the rejection of any "end-of-history"-connotations as included in the Marxist ideas, or even in the "end-of-utopia" discussion after the collapse of the socialist systems which were assuming that a certain "final state" of history should be or has already been reached. Instead, due to the standards of cultural rationality, the openness of history is to be emphasised. Future generations are fully responsible for their own affairs. The function of long-term considerations is not to determine the far future in the sense of envisaging a "final state" of history (this again would be

paternalistic) but to deliver orientation for present acting and deciding.[2] This structure of argumentation consists of a certain "deviation": *Presently* we have to act and to decide on technology affairs. We use long-term considerations on *future* expectations to get orientation for the *present* by "backcasting" (Vergragt and Janssen 1993). Forecasting (including normative reflection) and backcasting, therefore, are closely interrelated and both essential parts of the game (cf. Goldemberg *et al.* 1985). In the case of sustainable development, an appropriate guideline may be that visions concern a time span long enough to guide one generation to act so as not to burden the next generation (following this reasoning, Dutch environmental policy defines responsibility with a time horizon of 25 years).

3. Permanent reflection on the visions followed is indispensable: reflection on the normative premises of that vision, on the actual state of knowledge, on the purposes to be reached and on the interpretation of the relevant contextual aspects. Reflection on the goals and the means implemented to arrive at the goals may lead to (incremental) changes of direction in the development, of the goals as well as of the measures to reach the goals. This mechanism shall avoid the danger of running into a) misleading developments and persisting in them in spite of better knowledge and b) authoritarian biases in following guiding visions. From Michael Decker's chapter we can learn that a better understanding of how this can be done, even when visions are still embryonic, is one of the most urgent issues concerning vision assessment that requires further exploration.

4. Any assumption rooted in a "planning euphoria" – presupposing that future states of society could be planfully and expectably created – has to be rejected. As Grunwald (1999) has argued elsewhere, the position of planning euphoria is philosophically untenable. Regarding the future from the standpoint of *shaping* should not presuppose that pre-figured and fixed future conditions could be foreseeably created. Instead, shaping future technology constitutes an activity *involving risk* (enforcing the need for accompanying reflection, see above).

5. *Decentral* constitution and *decentral* modification of visions instead of central planning procedures taking into account the pluriformity of society and culture (Schwarz and Thompson 1990) and the perspectives of the stakeholders (compare the chapter of Rob Reuzel and Gert Jan van der Wilt) shall prevent technology policy from running into the well-known trap of an authoritarian utopianism. Utopianism in the former sense took the state as a central planning agent safeguarding the "correct" directions of development. This role, however, can-

---

[2] Compare the chapter of Michael Decker: the task of TA in the field of robotics is not to decide whether robots shall replace humans in the health care services 20 or 40 years in the future. Rather. it should achieve orientation with respect to the "next steps" of development, concerning, for example, demands for regulation, recommendations for promotion programmes of technology development or recommendations to perform public discourses.

not be fulfilled by the state in the modern pluralistic and highly differentiated society (Gottschalk and Elstner 1997).

6. The requirements for contextualisation in technology assessment (see the chapters of John Grin and Armin Grunwald) instead of applying decontextualised concepts to the problems arising also seem to be an important factor in order to prevent from following blindly some highly abstract utopias. On the other hand, it is not possible to resolve the ongoing affairs of society completely into mere casuistic contextualisation. In this way, the coherence of culture could be lost in favour of an atomistic society. Decontextualisation is indispensable in society, for example, if *societal learning* is aimed at: learning means to transfer the experiences reached in a certain situation to a different situation which requires some ideas of the "common" of both situations to justify the validity of the transfer (Habermas 1988).

7. As we have seen, the legitimation of long-term-related and normative ideas on the future of society has, ultimately, to be ensured by democratic decision-making procedures. Though such visions are often not explicit in the way that people could vote directly pro or contra the impact of them should be made as transparent as possible (for example by uncovering underlying assumptions, see below). In this way, there is a mechanism of democratic control on the role of visions in technology policy preventing again from blindly following them.

Together, the above points summarise the model of shaping technology through some form of directed incrementalism, in line with the concept of pragmatic rationality. The model seems to be very suitable to serve as a model for shaping technology by many small but reflected steps (compare also Rip et al. 1995; Schwarz and Thompson 1990; Grin and van de Graaf 1996a). It allows long-term directions to be maintained though the method of proceeding is the incremental one allowing short-ranged flexibility requirements to be taken into account. In this way, shaping the future is seen as making many small and reflected decisions in order to reach aims agreed upon under guiding visions. Each of these steps has to be processed under the reflection and control mechanisms mentioned above. To be as critical as we assume it should be, reflection has to go into the level of underlying assumptions. Let us now turn to the question where we might find and how we might surface these.

3   Uncovering Assumptions Underlying Visions

As we have seen, for both moral and pragmatic reasons we should not take underlying assumptions for granted, but, contrarily, uncover and discuss them. But how to uncover underlying assumptions? In fact, this amounts to two questions. The first, to paraphrase an intriguing question by Dvora Yanow (1993; 1996), reads: how does a vision assume?[3] The second question is: once we know where to look for underlying assumptions, how do we make these assumptions explicit? This section attempts to provide some answers to the first question and discusses some methods and techniques for dealing with the second.

Peter Mambrey and August Tepper's answer to the first question - how does a vision assume? - is: through empirically identifiable metaphors. Metaphors, here, should be understood as core concepts that are transferred from one artefact to another, analogous to paradigms in Kuhn's account of scientific progress. One might define metaphors as articulated similarities. We have, for example, the metaphor of a ten-finger operated machine that has been transferred from the piano to the typewriter ('Schreibklavier').

Now, these metaphors may include many assumptions regarding what the artefact should be. In order to, secondly, uncover these assumptions, the metaphors can be analysed in terms of discourses in areas related to the technology. For example, three discourses have been identified that are particularly relevant for the vision on the computer assistant for offices: those concerning organisation, those concerning co-operation and co-ordination, and those concerning man-machine interface. Aspects that appear relevant from these discourses concern the relation between technology, human behaviour and organisational routines; measures of effectiveness of organisation and management etc.

For instance, the discourse about man-machine interface, which is related to the master-slave metaphor, has steered the development of the office computer as an assistant to be operated by a human. According to Peter Mambrey and August Tepper, however, it is not only possible to explain artefacts in terms of metaphors and associated discourses, but also the other way round, to explain metaphors in terms of (future) artefacts. So, analysing metaphors enables us to enter debates about technology development at a very early stage, that is, when artefacts are not yet fully developed. Here, quite paradoxically, it seems that it is possible to profit from the fact that metaphors, in Fischer's (1980; 1995) theory, belong to the 'tacit' second order notions (see Introduction). For this makes that metaphors are fairly unsusceptible to sudden changes and can be used for future assessment.

Rob Reuzel and Gert Jan van der Wilt, rather than on metaphors, focus on artefacts, cochlear implants for example, as embodying visions, perspectives, and expectations. That is, they look at artefacts as being developed, implemented, used, and assessed with particular aims and expectations. This reflects Fischer's first order level. Visions assume when different perspectives and expectations require different parameters for development, use, and assessment. These visions are always there, of course, but particularly surface in case of conflict. Conflicts that may possibly arise form the perfect reason to scrutinise underlying assumptions, and to link first to second order notions. Such conflicts typically occur when there is, as in the case of cochlear implants, a strong relation between the artefact and its user's identity, and when effectiveness is judged differently by different stakeholders.

---

[3] In her work, Yanow has drawn attention to the fact that if it is of interest to ask 'what a policy means,' it is obviously also necessary to know 'how a policy means.'

Rob Reuzel and Gert Jan van der Wilt argue that uncovering underlying assumptions could be improved if the following suggestions are kept in mind. First, evaluators should use their imagination, and never regard any choice considering the aim and the course of an assessment trivial. Second, perspectives of different stakeholders should be actively sought for, and made explicit. Third, interaction between stakeholders could be established as to scrutinise the different perspectives, and try to achieve agreement about the conditions of the technology being acceptable to all stakeholders.

In John Grin's chapter, assumptions relate to the same kind of issues. Specifically, Grin focuses on the relation between strategic objectives, operational circumstances (human capabilities, organisational procedures) and technological systems; indicators for military effectiveness; as well as the relation between the military and other sources of power or influence of nation states or multi- or supra-national entities.

Armin Grunwald asserts that assumptions are related to value-orientations of both individuals and society as a whole. Moreover, he observes that these value-orientations represent different interests and are likely to conflict, especially when the trade-off between 'long-term planning' and 'short-ranged acceptance' is at issue. Very importantly, Grunwald sees that this problem is not so easily dealt with in pluralistic and democratic societies, in which any value-orientation has a basic legitimacy simply for its being held by someone, and cannot be overruled without being respected as such. This is even more true in the type of 'deliberative democracy' that is most prone to the type of endeavour we are interested in here. As Gutmann and Thompson (1996) have outlined in their political theory of deliberative democracy, 'reciprocity' (respecting each other as moral beings) is a key prerequisite for deliberation.

In order to ameliorate the paralysis that threatens to seize democratic societies accordingly, Grunwald proposes to take refuge in a *procedure* of dealing with technology development. To many people, such a procedure could be more acceptable, or is more easily accepted, than socio-technological developments themselves. For technological artefacts, having effects that are immediately visible, are much more prone to short-sighted rejection or embrace. Therefore, by focussing on a 'pragmatic rational' procedure, Grunwald attempts to simultaneously respect the visions of people affected by technology, and shield society from non-constructive responses and self-interests.

Michael Decker, too, asserts that there are assumptions, much in the line of Grunwald. Like Peter Mambrey, he assumes that already before the development of future artefacts the discussions about them starts and that is when the developers shape their perspective on the new technology. This should entail the first stage of technology assessment. Michael Decker argues that the assumptions underlying this first perspective can be made explicit in multidisciplinary expert panels. In these panels, experts deal with all kinds of issues: technical, economic, legal, moral etc. and therefore they are a promising way of scrutinising multiple perspectives in scientific and cultural rationality. Especially when the discussion in society has not started yet and long term orientation and a procedure of dealing with technology development are already needed. The results of such an expert discussion could be a good basis for wider public judgement.

In sum, in terms of the two questions this section started with, we have learned a twofold set of lessons. First, that we can best look for assumptions in those 'objects' that are, somehow, empirically identifiable. They may be the "policy objects" (Schön and Rein 1994) or "organisational artefacts" (Yanow 1993) 'through which visions assume'. In cases in which technological artefacts are still in an embryonic stage, these may be the metaphors expressed and used by those involved in their conception and development. In cases of artefacts already taking shape concretely, or starting to enter their application context, artefacts themselves may be the objects to assess.

Second, as regards the tasks and procedures that that may be suitable to uncover and discuss basic assumptions, it seems that we have been able to identify some important issues:

- Identify stakeholders. This is one of the basic questions in technology assessment, and one that up to now has not been resolved satisfactorily.
- Identify experts. What is meant here, is a broad range of experts who are able to shed light on a particular technology from various angles, and not the limited set of experts who are directly involved in developing and using technology.
- Discuss standards of performance and criteria of merit, as well as methods to assess these standards and criteria. Thus one sheds lights on various dimensions of the problem definition. Different visions possibly entail different problem definitions and different ways to assess proposed solutions. Looking for the assumptions behind them amounts to uncovering the empirical and normative assumptions that gave rise to that particular way to define the problem.
- Comparing objects central to the vision with alternative ones, perhaps with alternatives that fit more naturally in alternative visions, is a way to get stakeholder perspectives to the surface. This of course refers to the empirical and normative assumptions from which stakeholders look upon reality and set the problem. Visions may fundamentally differ from each other on that level.
- Discuss the relations between a technological artefact and the human and procedural factors that are envisaged to form its application context. This, of course, uncovers assumptions concerning the degree to, and the conditions under, which technology may contribute to solving societal problems. It is a key issue in the debate between true modernists and their critics of various kinds (cp. the introductory paper).
- Identify the basic features of the 'desired final state.' For example, what makes the security of nation states and trans- and multinational entities, and what is the precise definition of 'security'? What is the relation between various determinants of health and illness, and what constitutes health? What is the relation between technology, culture and structure as determinants of organisational effectiveness? This touches upon very basic assumptions around which visions develop: the preferred state of affairs the vision entails (Fischer's normative-ontological level, cf. the table in the introductory chapter).

4   Critically and Constructively Assessing Visions

Now that we know where and how to find the basic assumptions underlying visions, we have done all the prerequisite work to turn, in this section, to our main question: how to critically and constructively assess visions? Critically, here, refers not to any particular political stance that serves as an Archimedal point for review, but rather to the idea of assessing visions in terms of their basic assumptions. Constructively refers to what we have designated previously as a 'fusion of visions.'

Each of the previous chapters has brought us various methods for critical assessment in such terms. We will review them here, starting from the empirical field for which they have been illustrated before, but suggesting how their use may be extended into different fields as well. Thus this section will present to the reader part of a repertoire of methods for critically and constructively assessing visions.

Peter Mambrey presented discourse analysis as a method to assess metaphors embodying visions in terms of basic assumptions. For the assistant computer metaphor, these were discourses on organisation, on co-operation and co-ordination and on man-machine interfaces. These discourses, identified through some reflection on the substance and nature of the metaphor, were briefly recapitulated. Some key issues from each discourse were indicated and substantively filled in. Then the metaphor of the 'computer assistant' was assessed in terms of its relation to these key elements. Thus it was possible to judge to what extent the various discourses had shaped the metaphor. At least implicitly, this helps to identify potential alternative metaphors.

In this way, the method of discourse analysis contributes to critical assessment – it shows that metaphors are less self-evident than they might appear. And, although Peter Mambrey did not use it that way, it might also be used for constructing alternative metaphors. At least, it yields some alternatives. At best, an understanding of the discourses and their mutual relations may help to identify 'meta-metaphors' in order to construct visions that transcend existing, less plural, ones.

An interesting question is to what extent the concept of metaphors may be used to enable assessment of still embryonic visions, like that discussed in Michael Decker's chapter. Examples have been documented (e.g. Kensing and Madsen 1991) in which metaphors were used to present lay people with a clue of the roles that might be played by a technology that, hitherto, was a 'strange animal' to them. If this offers, indeed, some clues, it would be advisable in such cases to focus less on specific examples of technologies than on the metaphors implied in them. The interesting thing about the grid analysis employed by Michael Decker is that it both offers a typology of applications and critical features in terms of which they may be assessed, all from a rather straightforward reflection on the substance of the (types of) technology application. This grid can be used as a basis for the interdisciplinary discussion of experts. The proposal is that, by taking the perspectives of all disciplines found to be relevant into account, it is possible to assess existing visions critically as well as to combine them constructively into a more balanced vision. During this construction process all perspectives will be critically assessed and the arguments that sustain can be turned into (normative) recommendations. Especially for questions concerning long-term orientation for R&D these expert discussions are promising.

John Grin chose a different kind of empirical object in order to assess visions for 21st century warfare. Following an earlier developed method (Grin 1990; 1992; cf. Brauch *et al.* 1997, pp 81-85) he focused on military systems central to a particular vision, that could be more or less seen as a 'point representation' of at least some key elements of the vision. He then assessed that system within the context of one particular vision, in terms of criteria derived from both that vision and at least one alternative one. Thus fundamentally new 'guiding principles' (Smit 1989; 1991) may emerge from introducing a critical discursive moment (cp. section 1) in existing practices.

For instance, weapons for 'mass protection' are assessed in the framework of operational concepts for 'bloodless war,' in terms of their effectiveness of preventing 'unintended casualties.' The effectiveness criterion is operationalised in a way that takes into account the criticism against such typically modern visions: their claimed over-relying on technology, and de-emphasising operational limitations. Assessing the effectiveness of some exemplary weapons of mass destruction in the context of plausible operational conditions does this. As we have seen, different types of operational contexts may be envisaged: from the flat and easy terrain in Iraq up to, may be, the rain forest in Latin America. The relevant dimensions on which these contexts vary concerns, of course, the degree to and the ways in which one may expect operational conditions to have an impact on the effectiveness of the application of the military system under scrutiny.

In constructing a different vision, reflecting a wider set of normative and empirical assumptions into account, John Grin seemed to basically rely on analytical insight. First, he used his understanding on the relation between the analyses in various contexts; and second, he applied the heuristic of looking for visions that are about shaping the context rather than about relying on generally unwarranted assumptions concerning the relevance of the context.

In fact, while John Grin's method may be more explicit than Peter Mambrey's on the question how to assess a vision in terms of criteria reflecting a different vision's assumptions, it seems that applying his method presupposes some form of discourse analysis. Grin could do this without a form of explicit discourse analysis simply because he could rely on his understanding of the various discourses (Brauch *et al.* 1997, pp 82-84). That is, he did implicitly what Mambrey had made explicit. It is obvious that a more explicit treatment has the advantage of more transparency to the reader, and therefore of enabling more critical review by her or him. A minimum requirement would be to refer to the explicit reviews of these discourses that form the basis of this assessment.

On the other hand, John Grin actually went into the constructive exercise which Peter Mambrey actually stopped short of. Seen this way, the two methods discussed in the chapters by Mambrey and Grin, seem to be useful complements to each other. Yet, the unsatisfactory thing about both approaches remains the central role attributed to the analyst. Obviously, more interactive forms of analysis would do more justice to the fact that discourses are carried by *communities*. Actors participating in discourses do not just subsume reality under 'their' discourse, but rather construct, contextually, both the way in which discourses are to be made relevant and the light they shed on the object under scrutiny. This fundamentally reflective nature of the process of judgement makes a sort of *in vitro* simulation of judging artefacts or metaphors from the perspectives into a fundamentally risky undertaking. That is true for critically considering objects; it may be even more problematic in constructing new visions.

Against this background, the merits of Rob Reuzel and Gert Jan van der Wilt's argument for more interactive forms of assessment, especially for balanced visions and judgements on treatment methods, are obviously clear. To be sure, such methods do not make analytical activity from the side of the technology assessor superfluous. On the contrary, in actual examples of interactive technology assessment, analysts' interventions frequently appear indispensable to keep the process going (Hamstra 1993; Grin 1998; Grin et al. 1997, chapter 4). In that sense, the above ways to construct 'meta-metaphors' and 'meta-visions' may be of significant use in such interactively performed TA. The fundamental point about interactive TA practices is that interactions provide both the input and the validity test of any such intervention.

Two problems are common to all methods. First, introducing a critical discursive moment, if it is to have some impact on recursive practices, requires not merely methods, but also appropriate institutional fora. While this aspect is too important to be neglected, it goes way beyond the scope of this volume.

The second problem has been touched upon more directly: all methods discussed assume that, in addition to the dominant vision and corresponding discourses, sufficiently elaborated alternatives are available. In the case of the computer assistant and robotics the problem seems to be less the lack of alternative frameworks: organisation theory, communication studies and so on all contain a variety of theoretical paradigms. The main question concerns the knowledgeability of technology developers of these insights, or their tendency to taken them into account.

In some areas the problem may be tougher. Rob Reuzel and Gert Jan van der Wilt had to take recourse to deaf people's' stories in order to identify an alternative perspective, and they found that such perspectives hitherto lack similar elaboration as the more established one. John Grin pointed to the fact that only a handful of experts exist who have developed alternative approaches; embarrassingly enough, this implies that he has to rely on his own work to a significant extent. In contrast to this *ex post* analysis Decker claimed the necessity of an expert discussion *ex ante* in order to get a long-term orientation which makes directed incrementalism possible. Everybody is at liberty to scrutinise the results of the rational discussion afterwards. This may be even stronger the case in the military field, as the experiences with attempts to broaden central decision making routines (cf. section 1) suggests.

In areas like these, it seems, we can only conclude that the recently emerging idea of 'knowledge policy' is of paramount importance: our societies should, more deliberately and more actively than until now, organise pluriformity in its 'knowledge infrastructure.' That is, in these areas, deliberate governmental action may be a pre-requisite for mitigating the consequences of a historical structuration process that all too one-sidedly favours a particular set of assumptions.

So, contrary to some fashionable ideas, we arrive at the conclusion that the post-modern condition does not just imply that governments are facing a *mer-à-boire* of different standpoints, bothering any form of effective policy making. In order to make policies in more prudent ways, it may be necessary to make our societies truly more pluriform, through reinforcing the knowledge bases of non-traditional approaches. Such deliberate promotion of true plurality of knowledge infrastuctures appears a *sine qua non* for reaching more balanced visions to guide our entry in 21st century society.

References

Bijker W, Law J (1994, eds) Shaping Technology Building Society. MIT Press.

Brauch HG, van de Graaf H, Grin J, Smit W (1997) Militärtechnikfolgenabschätzung und Präventive Rüstungskontrolle. Institutionen, Verfahren und Instrumente. LIT Verlag, Münster

Braybrooke D, Lindblom CE (1963) A Strategy of Decision. New York

Constant, EW (1978) On the diversity and co-evolution of technological multiples: Steam turbines and Pelton water wheels, Social Studies of Science, vol. 8, pp 183-210

Dierkes M, Hoffmann U, Marz L (1992) Leitbild und Technik. Zur Entstehung und Steuerung technischer Innovationen. Edition Sigma, Berlin

Enserink, B (1993) Influencing Military Technological Innovation. Eburon, Delft

Fischer, F (1980) Politics, values and public policy: the problem of methodology. Westview press, Boulder, Col.

Fischer, F (1995). Evaluating public policy. Nelson-Hall Publishers, Chicago.

Fox ChJ, Miller HT (1996) Postmodern public administration. Toward discourse, SAGE Publications, London etc.

Giddens A (1984) The constitution of society. Outline of the theory of structuration. Polity Press, Cambridge

Goldemberg J, Johansson TB, Reddy AKN, Williams RH (1985) An end-use oriented global energy strategy, Annual Review of Energy 10, pp. 613-688

Gottschalk N, Elstner M (1997) Technik und Politik. Überlegungen zu einer innovativen Technikgestaltung. In: Elstner M (ed) Gentechnik, Ethik und Gesellschaft. Heidelberg, pp 143-180

Grin J, van de Graaf H (1996a) Technology Assessment as learning. Science, Technology and Human Values 20, no. 1, pp 72-99

Grin J, van de Graaf H, Hoppe R (1997) Technology Assessment through Interaction, Amsterdam

Grin J (1998) Participation, co-production and power. Rationale and praxis of interactively performed Technology Assessment: the example of the GIDEON project, Paper prepared for session B-1 (organised by Leeuw F and van der Wilt G), Participatory evaluation vs. top down evaluation, at The International Conference Evaluation: Profession, Business or Politics? European Evaluation Society, Rome, 29-31 October 1998

Grunwald A (1996) Die Bewältigung von Technikkonflikten. Theoretische Möglichkeit und praktische Relevanz einer Ethik der Technik in der Moderne, Zeitschrift für philosophische Forschung 51, pp 437-452

Grunwald A (1998) Technisches Handeln und seine Resultate. Prolegomena zu einer kulturalistischen Technikphilosophie. In: Hartmann D, Janich P (eds): Die kulturalistische Wende. Suhrkamp, Frankfurt, pp 178 - 224

Grunwald A (1999a) Technology Assessment or Ethics of Technology? Reflections on Technology Development between Social Sciences and Philosophy. Ethical Perspectives (to appear in October 1999)

Grunwald A (1999b) Handeln und Planen. Philosophische Planungstheorie als handlungstheoretische Rekonstruktion. Bouvier, Bonn

Grunwald A, Langenbach C (1998) Die Prognose von Technikfolgen. Methodische Grundlagen und Verfahren. In: Grunwald A (ed) Rationale Technikfolgenbeurteilung. Methodische Grundlagen und Verfahren. Springer, Berlin Heidelberg New York, pp 93-131

Habermas J (1988) Theorie des kommunikativen Handelns. Suhrkamp, Frankfurt

Hamstra AM (1993) Consumer acceptance of food biotechnology. The relation between product evaluation and acceptance. Part II: the method.. SWOKA, Den Haag

Harig H, Langenbach C (1999, eds) Neue Materialien für innovative Produkte. Springer, Heidelberg Berlin New York

Hartmann D, Janich P (1996) Methodischer Kulturalismus. In: Hartmann D, Janich P (eds) Methodischer Kulturalismus. Zwischen Naturalismus und Postmoderne. Suhrkamp, Frankfurt

Hatchett RL, Reuter RL (1992) Weapons research, development and acquisition in the United States, in: Smit WA, Grin J, Voronkov L (eds) Military-technological innovation and stability in a changing world. Politically assessing and influencing weapon innovation and military research and development,. VU University Press, Amsterdam, pp 85-94

Hoppe R, Grin J (1995) Technology Assessment for Participation: Experiences and Lessons, Industrial and Environmental Crisis Quaterly 9, pp 3-12.

Kensing F, Madsen KH (1991) Generating visions: future workshops and metaphorical design. In Greenbaum J, Kyng M (eds) Design at work. Cooperative design of computer systems. Lawrence Erlbaum Associates, Hilsdale, N.J. & Hove and London

Mambrey P, Pateau M, Tepper A (1995) Technikentwicklung durch Leitbilder. Neue Steuerungs- und Bewertungsinstrumente. Campus Verlag/St.Martin's Press, Frankfurt/New York

Morgan G (1997) Images of Organization. SAGE, London etc.

Paschen H (1999) Technikfolgenabschätzung in Deutschland – Aufgaben und Herausforderungen. In: Coenen R, Petermann T (eds) 25 Jahre Technikfolgenabschätzung in Deutschland. Campus, Frankfurt, pp 47-62

Popper K (1969): The poverty of Historicism. Routledge and Kegan Paul, London.

Rip A, Misa T, Schot J (1995, eds) Managing Technology in Society. London

Schön DA, Rein M (1994) Frame Reflection. Towards Resolution of Intractable Policy Controversies. The Free Press, New York

Schwarz M, Thompson M (1990) Divided We Stand. Harvester Wheatsheaf Press, Hassocks

Smit WA (1989) Defence Technology Assessment and the control of emerging technologies, in: Ter Borg M and Smit WA (eds) Non-provocative defence as a principle for arms reduction and its implications for assessing defence technology. VU University Press, Amsterdam, pp 61-76

Smit WA (1991) Steering the Process of Military Technological Innovation. Defence Analysis 7, no. 4, pp 401-415

van de Graaf H, Grin J (1999) Policy analysis and the 'reinvention of politics:' the question of closure. Paper presented at the Symposium on Policy, Theiry and Societ, Leyden, June 24-25.

van der Wilt G (1995) Alternative Ways of Framing Parkinsons Disease: Implications for Priorities for Health Care and Biomedical Research. Industrial & Environmental Crisis Quarterly 9, no. 1, pp 13-48

van Eijndhoven JCM (1995) De ondraaglijke lichtheid van het debat. De bijdrage van Technology Assessment aan het debat over wetenschap en technologie. University of Utrecht: Inaugural Lecture.

van Eijndhoven JCM (1997) Technology Assessment: product or process? Technological Forecasting and Social Change 54, pp 269-286

van Houwelingen J (1992) Politically influencing military technology: a policy maker's experience. In: Smit WA, Grin J, Voronkov L (eds) Military-technological innovation and stability in a changing world. Politically assessing and influencing weapon innovation and military research and development. VU University Press, Amsterdam, pp 125-132

Vergragt PJ, Jansen LA (1993) Sustainable technological development; The making of a Dutch long term oriented technology program. Project Appraisal 8(3) pp 134-140

Weyer J, Kirchner U, Riedl L, Schmidt JFK (1997) Technik, die Gesellschaft schafft. Soziale Netzwerke als Ort der Technikgenese. Edition Sigma, Berlin

Yanow D (1993) The communication of policy meanings: Implementation as interpretation and text. Policy Sciences 26, pp 41-61

Yanow D (1996) How does a policy mean? Interpreting policy and organizational actions. Georgetown University Press, Washington, D.C.

List of Authors

Decker, Michael, Dr. rer. nat., studied physics (minor subject economics) at the university of Heidelberg, 1992 diploma, 1995 doctorate at the university of Heidelberg, 1995-1997 scientist at the German Aerospace Center (DLR) in Stuttgart, since February 1997 member of the scientific staff of the European Academy, projectmanager of the project group "Robotics. Options of the replaceability of human beings". Main research areas: Technology Assessment, Comparison of TA-Concepts, Laserdiagnostics in High Pressure Combustion Processes. Address: European Academy GmbH, Postfach 1460, D-53459 Bad Neuenahr-Ahrweiler

Grin, John, Dr., MSc degree in physics at the VU University in Amsterdam 1986, Ph.D. 1990 at the same university on a thesis exploring the method of comparative defence technology assessment on technologies for command and control. Similar research since then at VU University, Dept. of Science and Society and the Arms Control Group at Princeton University. Since 1992 senior researcher at the Department of Political Science of the University of Amsterdam. Main research areas: sustainable technology development as a policy field, technology and security policy, (interactive) technology assessment, policy instruments and implementation. Address: Afdeling Politicologie, Universiteit van Amsterdam, OZ Achterburgwal 237, NL-1012 DL Amsterdam

Grunwald, Armin, Priv.-Doz. Dr. rer. nat., studied physics at the universities Münster and Cologne, 1984 diploma, 1987 dissertation on thermal transport processes in semiconductors at Cologne university, 1987-1991 systems specialist, studies of mathematics and philosophy at Cologne university, 1992 graduate (Staatsexamen), 1991-1995 scientist at the DLR (German Aerospace Center) in the field of technology assessment, since 1996 vice director of the European Academy, 1998 habilitation at the faculty of social sciences and philosophy at Marburg university with a study on culturalistic planning theory. Since October 1999 director of the institute of technology assessment and systems analysis (ITAS) at the research center Karlsruhe. Address: Institut für Technikfolgenabschätzung und Systemanalyse, Forschungszentrum Karlsruhe, Postfach 36 40, D-76021 Karlsruhe

Mambrey, Peter, Dr. sc. pol., M.A., studied Political Science, Sociology and Ethnology in Bonn and Duisburg. Senior Researcher at the GMD – German National Research Center for Information Technology, Institute for Applied Information Technology. Main research areas: Participatory System Design; Technology Assessment, Computer Supported Cooperative Work, and Community Network Analysis. Address: GMD, Schloß Birlinghoven, D-53754 Sankt Augustin

Reuzel, Rob, ir., studied Philosophy of Science, Technology and Society, University of Twente, Enschede, the Netherlands, on the basis of Applied Physics. Researcher at the Dept. of Medical Technology Assessment, University of Nijmegen, the Netherlands, since May 1996. Main research areas: ethics in health technology assessment, constructive technology assessment, and interactive evaluation procedures. Address: University of Nijmegen, Dept. of Medical Technology Assessment, 152 MTA, P.O. Box 9101, NL-6500 HB Nijmegen

Tepper, August, Dr. rer. pol., degree in business sciences (University of Kassel) and industrial sociology (University of Berlin). Joined the GMD Forschungszentrum Informationstechnik GmbH, Sankt Augustin, in 1981. Manager of the project "Metaphors in Computer Science - Potentials and Risks" As a member of the executive staff of the GMD, today he is responsible for the research field "intelligent multi-media systems". Main reserach areas: technology assessment, multi media, technology transfer, research planning. Address: GMD, Schloß Birlinghoven, D-53754 Sankt Augustin

van der Wilt, Gert Jan, Dr., MSc degree in biology at the Vrije Universiteit, Amsterdam, in 1986, with majors in neurobiology, endocrinology (Leiden, School of Medicine), and pharmacology (Amsterdam, School of Medicine). In 1990 PhD at the same university on a thesis on neural network analysis, then post-doctorate research fellow at the Insitute of Ethics of the Vrije Universiteit, studying social implications of health care technology, including implantation of fetal neural tissue as a treatment of parkinsonism. Since 1996 director of the department of Medical Technology Assessment at the University of Nijmegen. Address: University of Nijmegen, Dept. of Medical Technology Assessment, 152 MTA, P.O. Box 9101, NL-6500 HB Nijmegen